# Biotechnology:
# The Biological Principles

*M. D. Trevan, S. Boffey,*
*K. H. Goulding and P. Stanbury*

**Open University Press**
*Milton Keynes*

Taylor & Francis
New York • Philadelphia

Co-published by:
Open University Press
Open University Educational Enterprises Limited
12 Cofferidge Close
Stony Stratford
Milton Keynes MK11 1BY, England

Taylor & Francis
publishing office
3 East 44th Street
New York, NY 10017
USA

sales office
242 Cherry Street
Philadelphia, PA 19106
USA

First Published 1987

**British Library Cataloguing in Publication Data**

Biotechnology: The Biological Principles.
  (Biotechnology series Open University Press)
  1. Biotechnology
  I. Trevan, Michael D.
  660'.6    TP248.2

  ISBN 0-335-15153-1

  ISBN 0-335-15150-7 Pbk

**Library of Congress Cataloging in Publication Data**

Biotechnology: The Biological Principles.
  Bibliography: p.
  Includes index.
  1. Biotechnology.   I. Trevan, Michael D.
TP248.2.B575   1987        660'.6        87-1953

ISBN 0-8448-1506-3

ISBN 0-8448-1507-1 (pbk.)

Text design by Clarke Williams
Typeset and printed by The Universities Press (Belfast) Ltd.
Printed in Great Britain

# Contents

# Foreword

As the authors of this text rightly point out, biotechnology is an activity that has engaged the human race for a very long time—five or six thousand years at least. It is only since the advent of 'genetic engineering' however, that it has been clearly recognised as a key operation in human living. In this it is a prime example of how an activity with which we are all familiar but which has sunk into the background can be brought back to the forefront of attention by a new and exciting discovery. The ensuing flurry of activity is likely to start a whole new series of developments within the older areas of the subject. In the present instance, genetic engineering is not only the cause of the present realisation of the widespread importance of biotechnology but is, as John Higgins has pointed out, itself 'a dramatic example of the difficulty of predicting those areas of basic science which are most likely to lead to important applications. The discovery of recombinant DNA technology is a consequence of the large amount of support given to research into molecular biology for more than forty years. Yet even as late as the late sixties, a common criticism of giving relatively so much support to this 'glamour' area of chemistry and biology was that nothing directly useful had come from it. Now, however, it is clear that it has lead to discoveries that will radically affect humanity.

In the late fifties, a few pioneers, encouraged by the development of large scale fermentation by the antibiotic manufacturers, began to establish large scale biochemistry—'Biochemical Engineering' as it was then called since the term 'Biotechnology' was not then in use. However, not much notice was taken of these efforts until the understanding of the genetics of microorganisms had reached a stage where this knowledge could be exploited commercially. Based upon the fundamental understanding of bacterial genetics which resulted from the work of Monod and backed up by research on DNA replication, protein synthesis, the genetic code, restriction enzymes and bacterial plasmids, there emerged in the course of forty years an entirely new outlook in the biological sciences and one fraught with problems—scientific, ethical and industrial. The excitement engendered by this revolution and the power of the techniques that emerged encouraged many of the leading workers in the field to oversell and industrialists to overestimate the immediate commercial potential of the discoveries. Chiefly lacking was a clear appreciation that the genetics was only the preliminary stage to the production of a commercial product and that the intermediate steps were not only difficult but were in many instances unknown.

This text is designed to assist students and industrial people who are having to or who will have to grapple with these areas of biotechnology. Most existing texts of 'biotechnology' so far have tended to be either highly specialised or general surveys of the field—'what is Biotechnology'? We have so far lacked an intermediate level book dealing with the main background science relevant to biotechnology and which is suitable for middle year undergraduates and workers in other fields who find their activities impinging on biotechnology. Because of the wide range of the potential subject matter, the authors have had to be selective. They have very sensibly concentrated on their own areas of expertise so that the text is reliable and their enthusiasm comes over—always an enlivening aspect of a book. This one is a winner and is surely going to be well received.

**Eric M. Crook**
*Formerly Professor of Biochemistry*
*in the Dept. of Biochemistry*
*at St. Bartholomew's Medical*
*School, London*

# Preface

Publishing would seem presently to many the most lucrative activity under the umbrella of biotechnology, so why yet another book? This particular book is written with students of biotechnology firmly in mind. It has arisen out of the author's own experience of teaching in the biological disciplines underpinning current developments in biotechnology, both at undergraduate and postgraduate level. We have aimed to write a book which takes the student on from his basic knowledge of biochemistry, microbiology and molecular biology and helps to develop that understanding of the rationale and problems in the specific areas of microbial metabolism, the growth and culturing of microrganisms, genetic manipulation and biocatalyst technology. Thus the book is divided into four major sections covering each of these topics. In subdividing the book in this way, we were aware of the multidisciplinary nature of biotechnology, but also keenly cogniscent of the need for an integrated approach and have thus attempted, wherever sensible, to cross-reference the text and to write the book as a coherent whole, ensuring an even level of treatment throughout. Time alone will tell how successful we have been. In order to achieve this, each section is prefaced by a list of concepts of which a basic understanding is assumed, together with a few suggestions of appropriate standard texts to which the reader may refer.

Thus we would have a continuous theme; microbial metabolism, how and why microbes grow and on what; how microbes can be cultured; the products which can be derived from them; how productivity can be improved by both conventional methods and genetic engineering; how genetic engineering is performed and what else it can do; the value of enzymes as a group of biological compounds which are both important microbial products and ameanable to the techniques of genetic engineering; how enzymes may be isolated and purified; how their catalytic activity may be harnessed and enhanced; and the uses to which enzymes may be put, including enhancing microbial productivity and in the techniques of genetic engineering.

To aid the readers' understanding, we have used pictorial and tabular presentation of the material as often as possible to supplement the text, and have also included brief summaries at the end of each chapter. We also hope that this book will prove useful to the practising biotechnologist who requires an introduction to a subject outside his own area of expertise.

In this particular presentation we have deliberately chosen to exclude two important biological aspects of biotechnology, plant cell and animal cell tissue culture. We have done this both because of the rather more specialist nature of these subjects, rarely tackled effectively at undergraduate level, and also in order to keep the text to a reasonable length.

More controversially, we have deliberately taken an intuitive approach to those areas which usually involve mathematical formulae, attempting to develop an innate understanding onto which can be grafted a detailed knowledge of the necessary mathematical manipulations. For reasons of clarity of the text, we have also chosen to leave out the EC. numbers for enzymes and have instead included a glossary of systematic enzyme names and numbers.

Finally we would like to record our thanks to all those of our friends and colleagues who have made helpful comments during the preparation of this book, to the editorial staff of Open University Press without whose encouragement and threats this book might never have been completed and to our families without whose sufferance and understanding this project would have been impossible. We hope that this book lives up to its aims; any mistakes or misconceptions remain the fault of the authors and we would welcome any comments or criticisms from our readers.

*M.D. Trevan*
*S.A. Boffey*
*K.H. Goulding*
*P.F. Stanbury*

# Figure acknowledgements

The following figures have been reproduced by kind permission of the copyright holders.

Figure 6.1 from Malik, V. (1980) *Trends in Biochemical sciences*, **5** (3), 68–72, pub. Elsevier/North Holland Biomedical Press, Amsterdam.

Figure 7.2a Butterworth, D. (1984) In *Biotechnology of Industrial Antibiotics*, Ed. Vandamme, E. J., pp 225–236, pub. Marcel Dekker, New York.

Figure 7.2b Podojil, M., Blumaverova, M., Culik, K. and Vanek, Z. Ibid, pp 259–280.

Figure 7.2c Okachi, R. and Nara, T. Ibid, pp 329–366.

Figure 8.2a Dawson, P. S. S. (1974) Biotechnology and Bioengineering Symposium **4**, 809–819 pub. John Wiley, New York.

Figure 8.2b Taylor, I. J. and Senior, P. J. (1978) Endeavour, **2**, 31–34. pub. Pergamon Journals Ltd., Oxford.

Figure 8.2c Smith, S. R. L. (1980) Phil Trans. Roy. Soc. (London) B., **290**, 341–354. pub. Royal Society, London.

Figure 8.2d Hamer, G. (1979) In *Economic Microbiology*, Vol. 2. Ed. Rose, A. H. PP 31–45. pub. Academic Press, London.

The following figures and tables have been reproduced by kind permission of the copyright holders.

Figure 13.3 from Bruton C. J. in *Industrial and Diagnostic Enzymes* (Phil. Trans. R. Soc. Land. B **300**, (1983)) 246–261 Ed. Hartley, B. S. *et al* pub. Royal Society, London.

Figure 14.6 from Katchalski E. & Goldstein, L. (1972) Biochemistry **11**, 4072, American Chemical Society.

Figures 14.7, 14.13, 14.17 and 14.19 from Trevan, M. D. (1980) *Immobilized Enzymes* pub. John Wiley & Sons, Chichester.

Figure 14.11 from Wharton, C. W., Crook, E. M. Brocklehurst, K. (1968), Eur. J. Biochem **6**, 572–578 pub. Federation of European Biochemical Societies.

Figure 14.15 from Simon, L. M. *et al* (1985) Enz. Microb. Technol. **7**, 357–360 pub. Butterworth & Co. (Publishers) Ltd.

Table 14.7 from Chibata I & Tosa, T. (1976) in Appl. Biochem. & Bioeng. **1**, Ed. Wingard H. L. *et al* pp 334–5 pub. Academic Press.

# SECTION I
# Introduction

# Chapter 1

# *What is biotechnology?*

'Superbugs for the third industrial revolution' is the kind of media headline, prevalent in recent years, that is guaranteed to send a chill down the spine of any biotechnologist. The problem is one of perspective. The common view of biotechnology is often distorted both in the appreciation of what it is and what it might achieve. It is the purpose of this brief introductory chapter to provide an overview of biotechnology, who (or what) biotechnologists are, and what biotechnology is worth.

## 1.1   What?

Biotechnology has been defined in various ways, mostly rather unsatisfactorily, but if pressed a biotechnologist would probably recite 'The application of biological organisms, systems or processes to manufacturing and service industries'. This rather vague definition in practice means 'biology applied for profit, either financial or, less often, humanitarian'. In fact biotechnology does not exist as a scientific discipline, nor is it an emerging interdisciplinary field, rather it is multidisciplinary, involving a wide variety of distinct subject areas. Indeed, such is the breadth of knowledge and skills encompassed by the term that a meeting of biotechnologists can be reminiscent of a scene from the Tower of Babel, such is the diversity of language and jargon employed.

The term 'biotechnology' was most recently brought into popular usage in the mid-1970s as a result of the increased potential for the application of the emerging techniques of molecular biology. The word itself first seems to have been employed by Leeds City Council in the United Kingdom in the early 1920s, when they set up an Institute of Biotechnology. In fact, biotechnological process precede even that date by some 5000 years with the discovery of the production of alcoholic beverages by fermentation.

The ancient Egyptians went several stages further than even this time-honoured art, with the use of mouldy bread as poultices for infected wounds (the forerunner of antibiotics), and the introduction of a pregnancy testing service based on the effect of urine on the germination rate of wheat and barley (but that story is sadly beyond the scope of this present text). Thus biotechnology can claim to be a modern technology which is as old as the hills! Present interest in biotechnology has been stimulated by the potential that can result from the marriage of biological processes and techniques—some old, some new—with production engineering and electronics. The fruits of biotechnology are born on a tree whose roots are the biological sciences, in particular microbiology, genetics, molecular biology and biochemistry, and whose trunk is chemical engineering in its widest sense. The purpose of this book is to serve as an introduction to the principles of the three major roots: microbiology, biological catalysis, and molecular biology.

## 1.2  Who?

It should by now be apparent that defining a biotechnologist is even more difficult than defining biotechnology, not just because of the wide range of disciplines involved, but also because of the transient nature of the identification paraded by an individual. One may be a biotechnologist one week and a biochemist (or microbiologist, etc.) the next. If we were to list all those who saw themselves making a contribution to successful biotechnology, the list would include biochemists, microbiologists, geneticists, molecular biologists, cell biologists, botanists, agricultural scientists, virologists, analytical chemists, biochemical engineers, chemical engineers, control engineers, electronic engineers and computer scientists. Even then the list would be incomplete for we must include economists, accountants, managers, that is, those who are responsible for creating a marketable technology from an interesting scientific pursuit. It is salutary to consider that all that is scientifically and technically possible is not necessarily worthwhile or profitable.

## 1.3  How many and where?

It is impossible to generalize on the size of the biotechnology community or where they work, but a few pointers can be provided. International scientific conferences on biotechnology attract upwards of 3000 participants and this figure probably represents less than 10% of the total community. In the United Kingdom alone it is estimated that, excluding traditional industries such as brewing, there are some 2000 biotechnologists and that this figure is likely to grow by 20% per annum for the next 10 years.

The employment distribution broken down by place of work (for example, industry, research institute, higher education institution) varies from country to country, but probably the minority are employed by industry, the chemical and pharmaceutical industries being the major employers, either directly or through the financing of small independent biotechnology companies. In the research institute sector the major preoccupations are with health and agriculture. *Omnipresence* is the key word in institutions of higher education.

## 1.4  What do they do?

Activity in biotechnology may be conveniently broken down into eight areas of endeavour, outlined in Fig. 1.1.

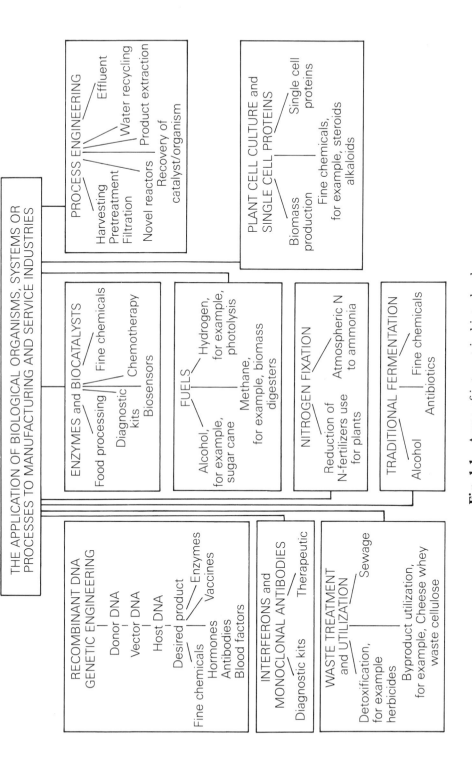

**Fig. 1.1**  Areas of interest in biotechnology

### 1.4.1 *Recombinant DNA and genetic engineering*

Of any single field, that of molecular biology has probably provided most of the impetus towards the unification of modern biotechnology. The ability to extract a gene coding for a desired product and transfer it to another organism has opened the way either to the more effective production of useful proteins, or to the introduction of novel characteristics in the host organism. Thus the large-scale production of hormones, vaccines, blood clotting factors or enzymes by some friendly bacterium becomes possible. But why go to all this trouble, why not just extract the desired protein from the original source? There are four reasons. First, it is often not practical to grow certain types of cell on a large scale. For example, mammalian cells, particularly of human origin, may be difficult to obtain, grow slowly, and are not amenable to the simple culturing techniques available for the growth of micro-organisms. It has been speculated that the production of interferon from cultured human cells is likely, in the longer term, to be supplanted by production from genetically engineered microorganisms. This, of course, presupposes that the differences in structure between interferons produced in these two ways has no significant effect on their action. Second, supply of the natural source material may be strictly limited. Media-induced public moral indignation is rife when stories come to light of mortuary attendants removing pituitary glands for the extraction of human growth hormone. Third, the natural supply may be unavoidably contaminated. Haemophiliacs receiving factor XIII isolated from serum have run the occupational risk of also receiving hepatitis or, latterly, AIDS. Fourth, is the cost.

There is also an additional possibility with this form of technology, the production of truly novel proteins. Let us take the example of enzymes. The use of enzymes as industrial catalysts is limited in part by the properties of those enzymes that are available. These properties, in particular specificity, catalytic ability and stability, are governed by the precise structure of the enzyme molecule. By selectively modifying the gene coding for the enzyme before it is introduced into the host organism, the structure, and hence properties of the enzyme, may be advantageously modified—hence, a new breed of superenzymes.

Modification of the genome of economically important plants also holds out great promise. The introduction into crops of the ability to fix nitrogen from the atmosphere would not only save on the cost of applying nitrogenous fertilizers, but would eliminate the potential problem of water pollution from nitrates washed off agricultural land. It is estimated that nitrogen-fixing brussel sprouts could be produced at half the cost of the traditional variety. Whether half-price brussel sprouts are a high priority is perhaps more questionable! The levels of storage proteins in seeds could be increased to give, for example, high protein wheat. It is possible that crops could be engineered for a greater resistance to herbicides or infection.

### 1.4.2 *Mammalian cell culture*

It is likely, for sound scientific or economic reasons, that some mammalian proteins can only be produced from cultured mammalian cells. The two most likely candidates in this category are monoclonal antibodies, for reasons of the complexity of the transcription and translation of their genetic material, and interferons, for reasons of cost and effectiveness. These classes of proteins are likely to become even more important in the future both for therapeutical preparative and analytical applications (see biosensors, Chapter 14, Section 14.7). Thus the problem of culturing mammalian cells on a large scale is likely to be a major preoccupation of cell biologists and biochemical engineers for the foreseeable future.

### 1.4.3 *Plants and plant cell culture*

Plants, quite apart from their key role in providing food, are also important sources of other raw materials. Possibly the most important, presently used bulk plant products are starches and sugar; 90% of cars in Brazil now run on a mixture of petrol and alcohol, the latter derived by fermentation of cane sugar. Sugar has also been proposed as a feedstock for the chemical industry, and the technology exists for its conversion into ethylene oxide. At present the price differential between sugar and ethylene (Table 1.1) is not great enough to make this a commercially viable proposition.

Plants are also an important source of high value drugs; some 25% of the drugs in a modern pharmacopoeia are of plant origin. However, deriving such high value compounds from cultivated plants involves reliance upon factors such as the weather, politics and market economics, which make the economics of production somewhat volatile. It is largely for these reasons that the science (some would say art) of plant cell culture has been developed. The ability to culture plant cells on a large scale, either for the production of biomass *per se*, or in order to extract the desired product from (high yielding) cell cultures, is becoming a highly desirable technology. Most of the valuable compounds thus produced are so called secondary metabolites, synthesized by the cell in its stationary phase (that is, when it is not growing actively). An extension to this technology is the immobilization of plant cells (see Chapter 14, Sections 14.2.3; 14.2.4.7; 14.4.) so that they can be held, at high concentration within a bioreactor, for months on end. If ever there was factory farming, this truly must be it! At present the technology is expensive, so potential products must be of high value (typically in excess of $500 per kg). However, as the technology advances and costs fall it may be economic to produce low cost, high volume materials, for example glycolic acid as a feedstock for the chemical industry.

**Table 1.1**  Bulk prices of biological and other products (1980)

| Biological | Price £ tonne$^{-1}$ | Non-biological |
|---|---|---|
| | 10 000 000 | Gold |
| Jasmine oil | 4 000 000 | |
| Saffron | 4 000 000 | |
| Vitamin B$_{12}$ | 2 300 000 | |
| Capsaicin (chilli) | 250 000 | |
| Penicillins | 45 000 | |
| | 5500 | Polytetrafluoroethylene |
| Cheese | 1300 | |
| | 1270 | Aspirin |
| Citric acid | 700 | |
| | 500 | Ethylene oxide |
| Yeast | 460 | |
| Sugar | 350 | |
| Beer (untaxed) | 280 | |
| | 160 | Naphtha |
| Soya bean meal | 115 | |
| | 85 | Crude oil |
| Molasses | 80 | |

### 1.4.4   *Fuels*

The world, so we are told, is running short of combustible fuels, in particular mineral oil, and as supplies begin to fail what oil there is will be too valuable as a feedstock for the chemical industries (for the production of plastics, etc.) to be burnt as a fuel. Biotechnology might provide two solutions, new fuels (see Chapter 15, Section 15.4) and alternative carbon feedstocks. The economics of the Brazilian fuel alcohol process (see above) are finely balanced, as the alcohol must be distilled before use, an energy-consuming process. In this case the process is made energetically economic (just) by using the sugar cane waste as a fuel for distillation.

Another potential biotechnological fuel is methane. All that is needed is some pig slurry (or something similar) and a hole in the ground with a lid on top; nature does the rest. Obviously there is room for improvement of such technology, but it is one example of biotechnology that can be transferred to third world agricultural societies without much difficulty.

Perhaps the most aesthetically pleasing biotechnological fuel would be hydrogen derived from the biophotolysis of water. The technology for storage and handling of hydrogen already exists, and in some respects it is safer in use than petrol and a lot lighter into the bargain. Biophotolytic production of hydrogen from water has already been achieved, and is based on the combination of the photosystems of plants with bacterial-derived hydrogenase enzymes and light. At this time a number of problems remain: the instability of hydrogenases; and the relative inefficiency of the conversion of light to energy by chlorophyll. Nevertheless reports exist of such systems remaining in operation for several months. The great virtue as a fuel of hydrogen derived from water is that, when burnt, it produces no pollution and regenerates its source material. In this respect at least it must be the ideal combustible fuel.

### 1.4.5   *Biocatalysis*

Enzymes are nature's supreme catalysts, exhibiting great specificity and enormous catalytic power. Their potential for a wide variety of applications is a major concern of this book. They have been in use for centuries, particularly in food processing (for example, cheese making, hair removal from hide) and represent one of the oldest forms of biotechnology. More recently, great interest has been aroused in extending the use of enzymes (whether purified, as dead or partial viable cells) in food processing, chemical production, analytical and diagnostic systems, and in the treatment of disease. This modern interest is largely a result of the greater understanding of the function and nature of enzymes and also the new technologies involved in the handling of biocatalysts, particularly their immobilization so that they may be fixed in a reactor or on a sensor probe.

### 1.4.6   *Waste treatment and utilization*

Sewage disposal is a problem long faced by mankind. Gone are the days when a cautionary shout from an upstairs room would be quickly followed by the contents of a chamber pot. Today's sewage plants are good examples of simple biotechnology: a fixed bed of microorganisms degrading the sewage waste products trickling over them. However, many other forms of waste exist which, with the appropriate technology, could not only be more easily disposed of, but perhaps turned into a useful commodity. For example, cheese production starts with the curdling of milk to form the solid curds and liquid whey. Unlike Little Ms Muffet we tend not to eat the whey but pour it down the drain. An average size cheese-making plant can produce thousands of litres of whey a day, and tipping it into the sewers is not only problematic but also costly. Whey is composed of a few proteins, minerals and about 4% lactose. The lactose as such is not particularly useful; two-thirds of the world's population cannot digest it; and it is neither very sweet nor soluble. However, it does find some application in the manufacture of ice-creams, 'packet' soups and desserts; but its use could be widely extended if the lactose were split into its constituent sugars, glucose and galactose. Processes have been developed around the use of the enzyme β-D-galactosidase to do just this.

Cellulose is another abundant waste material, particularly in the form of straw from cereal crops. Traditionally this is either ploughed back into the fields or, in recent times, burnt. The burning of straw is gradually being prohibited in the developed world, largely for environmental reasons. In theory, this waste cellulose could be biologically degraded and used as a feedstock for the production of microbial proteins. It has been estimated that sufficient protein could be produced in this way from agricultural waste alone to feed the entire world's population—a sobering thought.

Finally, there are those materials such as herbicides that once they have served their legitimate purpose are effectively problematic waste materials. Here too it may be possible to develop biological methods for their *in situ* detoxification.

### 1.4.7 *Fermentation*
Fermentation shares with biocatalysis the distinction of being the oldest form of biotechnology. Traditionally, fermentation has meant the production of potable alcohol from carbohydrates. However, fermentation—that is, the application of microbial metabolism to transform simple raw materials into valuable products—can produce an amazing range of useful substances, for example, chemicals such as citric acid, antibiotics, biopolymers and single cell proteins. The potential is as endless and varied as micro-organisms themselves. What is needed is the knowledge of these micro-organisms, the control of their metabolism and growth, and the ability to handle them on a large scale. These areas are dealt with at length in subsequent chapters of this book (Chapters 2–9).

### 1.4.8 *Process engineering*
It is one thing for the laboratory scientist to clone a novel gene, discover a new antibiotic or invent an enzyme catalysed process, but it is quite another to transfer this knowledge to the scale of operation required to make a useful product in significant quantity. This last is the remit of the engineer, be he chemical, biochemical or whatever. Harvesting, pretreatment, filtration of the raw material, reactor design and control, recovery/reuse of biocatalyst or organism, product extraction and analysis, effluent treatment, water recycling—all these are his concern. Chemical engineers have shown themselves to be very adept at handling chemical processes on a large scale. However, biological processes differ from chemical processes in many important ways, for example, the requirement for sterility or axenic production, or the extraction of labile materials diluted in large volumes of water. It is questions such as these, and others yet to be encountered, that present a great challenge to man's technological ingenuity. Without their solution the biologist might well pack up his technology and return to natural history.

## 1.5  How much?

What then is biotechnology worth? The answer to this question lies in both the inherent value of biologically derived products and the contribution that industries

related to biotechnology make to the gross domestic product (GDP) of an economy.
Table 1.1 draws an interesting comparison between the bulk prices of biological and
other products.

Top of the list is gold, but running a close second are jasmine oil and saffron.
Perhaps in an alternative culture these would be the hoarded national assets.
Examination of the relative prices of sugar and ethylene oxide has already been
suggested (see above). One of the most surprising comparisons though is probably
that of the costs of cheese and aspirin (roughly equivalent). In retail terms, however,
aspirin costs some 50 times more than cheese. This is not so much a reflection of the
relative profit margins of the food and pharmaceutical industries as an illustration of
the principle of the importance of market volume. The relative value of a product
depends not only on the price for which it can be sold, but also on how much of it can
be sold. This consideration is also illustrated by Table 1.2 giving the total world
market value of certain biological products; the production of jasmine oil, that

### Table 1.2   World markets for biological products (1981)

| Product | £ Million |
|---|---|
| *Alcoholic beverages | 23 000 |
| *Cheese | 14 000 |
| **Antibiotics | 4500 |
| **Diagnostic tests | 2000 |
| †Seeds | 1500 |
| *High fructose syrups | 800 |
| *Amino acids | 750 |
| *Yeast | 540 |
| **Steroids | 500 |
| **Vitamins | 330 |
| *Citric acid | 210 |
| *Enzymes | 200 |
| **Vaccines | 150 |
| **Human serum albumin | 125 |
| **Insulin | 100 |
| **Urokinase | 50 |
| **Factor VIII | 40 |
| **Growth hormone | 35 |
| †Microbial pesticides | 12 |
| **Jasmine oil | 8 |
| Total: | 48 850 |

### Summary

| Product | £ Million |
|---|---|
| * Foods | 39 500 |
| ** Health | 7830 |
| † Agriculture | 1512 |

**Table 1.3** Contribution of biotechnology-related industries to GDP (as % OEDC 1978)

| Country | Food | Industry Chemical | Health |
|---------|------|-------------------|--------|
| France | 10.5 | 12.8 | not available |
| Germany | 7.4 | 15.1 | 1.1 |
| Japan | 8.9 | 14.9 | 1.4 |
| New Zealand | 15.5 | 6.0 | 0.3 |
| Norway | 13.2 | 8.8 | 0.2 |
| Sweden | 7.9 | 7.1 | 0.5 |
| United Kingdom | 9.8 | 14.8 | 1.0 |
| United States | 9.1 | 13.6 | 0.8 |

exceedingly expensive commodity, is in fact a low value 'industry' (£8 000 000) because there is relatively little demand for it. Nevertheless, it can be clearly seen that biological products are valuable both in terms of bulk price and market value.

If industries related to biotechnology are classified according to the nature and mode of manufacture of their product we can see (Table 1.3) that, excluding agriculture, they make on average a 20–25% contribution to the GDP. In terms of national economic growth as a result of the introduction of biotechnology, however, it is clear from these figures that the greatest potential lies in the food and chemical industries, rather than the pharmaceutical/health care industries. It is perhaps surprising therefore that many a national strategy for biotechnological development is aimed at these self-same health care industries and that the majority of government funded research and development is in these fields (over 60% of total government funding in the United Kingdom). Whether this is related to human altruism, or to the fact that many of the products of biotechnology most immediately attainable happen to fall within the area of health care, is open to debate. How much different national governments spend on funding research and development in biotechnology is shown in Table 1.4, although care must be exercised in interpreting these figures.

**Table 1.4** Government funding of biotechnology research and development (1983)*

| Country | £ Million |
|---------|-----------|
| United States | 495 |
| West Germany | 96 |
| Japan | 90? |
| France | 66 |
| Britain | 60 |
| Italy | 27 |
| Netherlands | 21 |
| Belgium | 13 |

* EEC estimates.

**Table 1.5**   OECD predictions in biotechnology products from genetic engineering (1981)

| | |
|---|---|
| Human insulin | 1985 |
| Interferon (cancer) | 1987 |
| Interferon (antivirial) | 1988 |
| Hepatitis B vaccine | 1990 |
| Growth hormone | 1990 |

In conclusion then it can be said that biotechnology is already a valuable asset that future developments can only increase.

## 1.6   To what end?

We have already seen in the preceding sections of this chapter something of the scope and potential of biotechnology. The development of biotechnology will be a long-term affair, dependent on the whim of market forces and developments in competing technologies. Predicting the future is a fool's game at the best of times, but maybe certain foresights and some pitfalls can be usefully recorded. Table 1.5 is a list of OECD predictions, made in 1981, of when it was judged that certain products from genetic engineering would be commercially marketed. It is interesting for two reasons. First, it would appear at this time that these predictions are proving to be fairly accurate. Second, some of these predictions are being overtaken by other developments. Thus, chemically modified pig insulin, indistinguishable in molecular terms from human insulin, appeared on the market in 1983. Interferon, the wonder drug, may not be so wonderful after all; cancer treatment may be ultimately advanced not by interferon but by newer discoveries such as TNF (tumour necrosis factor).

## 1.7   Summary

The key to the successful development of biotechnology lies either in producing a product which can be made by no other means, or by producing an existing product more cheaply. Quite obviously, however, the latter approach involves catching up on

**Table 1.6**   Possible medium-term biotechnology products and services (from Dunnill and Rudd (1984))

Vaccines for common cold
Competitive (safer) tobacco substitute
'Single cell' collagen for steak
Cheap 'premier' wines
Reliable self-diagnosis kits
Novel flowering plants
Analytical methods to assess human response to new foods, drugs
Site-specific 'targeted' drugs

existing technology, which itself may be advancing, and thus involves considerable risk. Table 1.6 lists some of the more obscure biotechnology products and services which could be targets for medium-term development and which reflect the former of these approaches.

Is biotechnology to be the next industrial revolution? So much is uncertain, but we can be sure that biotechnology is here to stay.

# SECTION II
# Microbial growth

*K. H. Goulding*

**Assumed concepts**

Basic metabolism, such as glycolysis, TCA cycle, Fatty acid metabolism.
Energy conservation, such as oxidative phosphorylation, photophosphorylation.

*Useful books*
Lehninger, A. L. (1982). *Principles of Biochemistry*. New York, Worth Publishers.
Stryer, L. (1981). *Biochemistry*. San Francisco, W. H. Freeman and Co.

# Chapter 2

## *Introduction to metabolism*

It is often claimed that biochemistry is the unifying theme of the biological sciences. The evidence to support such a statement comes largely from comparative studies on cellular metabolism and energy conservation mechanisms. Apart from the fundamental structural and organizational differences between prokaryotic and eukaryotic organisms (which include, in prokaryotes, the lack of a nuclear membrane and organelles such as mitochondria and chloroplasts and differences in the organization, replication and expression of the genetic material), all organisms show marked similarities at the cellular level. In fact, even many structural features and aspects of the organization, replication and expression of the genetic material are remarkably similar in bacteria, higher plants and animals, but the similarities between these groups are particularly striking when basic metabolic processes are compared.

## 2.1 Generation of ATP

To grow and maintain themselves all organisms must be able to produce the high energy nucleotide adenosine triphosphate (ATP), in order that it can be used in the many energy-consuming processes, such as transport and biosynthesis, which are essential for life. Biosynthetic processes themselves depend on an adequate supply of carbon and nitrogen and generally start from a small pool of, so-called, *central metabolic intermediates*. Although there are variations, the pathways involved in ATP production are fundamentally similar in all organisms.

Plants, cyanobacteria and other photosynthetic bacteria produce ATP primarily as a result of photosynthetic electron transport mechanisms and are known as *phototrophs*. On the other hand in *chemotrophs*, ATP is derived from the oxidation of organic compounds which, in micro-organisms, are referred to as growth substrates. The oxidative pathways involved and the associated electron transport processes are remarkably similar in all organisms. In aerobes the process is usually referred to as *cellular respiration*, while in anaerobes, or in aerobes functioning under anaerobic conditions, it is referred to as *anaerobic respiration* or *fermentation*.

The vast majority of organisms use very similar respiratory mechanisms which are comprehensively discussed in most biochemistry text books (see, for example, Lehninger, 1982; Stryer, 1981) and are therefore not described in detail here. Thus only a basic outline of respiratory metabolism is shown in Fig. 2.1. This shows that virtually any type of growth substrate which contains three or more carbon atoms can be oxidized, directly or indirectly, via intermediates of the *glycolytic pathway* to acetyl CoA which is then completely oxidized to carbon dioxide via the *tricarboxylic acid cycle* (TCA or Kreb's cycle). Thus, substrates such as polysaccharides, monosaccharides and their phosphates, organic acids, and amino acids are oxidized to glycolytic intermediates which are then further oxidized to acetyl CoA while lipids are oxidized directly to acetyl CoA via such pathways as *β-oxidation*. Acetyl CoA then enters the TCA cycle by condensing with the four-carbon organic acid, oxaloacetic acid, to form the six-carbon citric acid. Citrate is oxidized via a cyclic series of reactions which result in the regeneration of oxaloacetate and the release of two molecules of carbon dioxide. Effectively, therefore the TCA cycle oxidizes the two-carbon acetate to two molecules of carbon dioxide. At various stages in these oxidative events, electron acceptors (X in Fig. 2.1) such as nicotinamide adenine dinucleotide ($NAD^+$) are reduced. These are subsequently reoxidized via an *electron transport chain* involving, among other intermediates, various cytochromes which are alternatively oxidized and reduced until the terminal step in which oxygen acts as the electron acceptor and water is produced. During these electron transport reactions energy is released some of which is conserved during the process of *oxidative phosphorylation*. This results in ATP generation.

Thus, with respect to the aerobic utilization of substrates with three or more carbon atoms, the process of ATP generation is very similar in all organisms. However, there are differences in micro-organisms, particularly those capable of growing anaerobically or at least tolerating anaerobic conditions or those capable of growing on substrates with less than three carbon atoms such as methane, methanol and glycollic acid. Since such organisms are extremely important in biotechnology these variations are considered fully in Chapters 3 and 4. In addition, and also because of their biotechnological importance, the pathways involved in the oxidation of some other compounds such as higher hydrocarbons and aromatic substrates, that involve special methods of converting the substrate to intermediates of glycolysis or the TCA cycle, are considered in Chapter 5.

## 2.2    Generation of biosynthetic precursors

The second requirement for growth and maintenance (ATP generation being the first) is the ability to synthesize cellular components such as amino acids (as precursors for proteins), monosaccharides, fatty acids and other lipids and the nucleotide precursors of nucleic acids. Again, in the majority of organisms, there is a remarkable similarity in how this requirement is achieved. Details of the pathways involved are discussed fully in most biochemistry textbooks, so again only a brief commentary is given here.

As shown in Fig. 2.1, most *biosynthetic routes* start from intermediates of either the glycolytic pathway or the TCA cycle. For instance most amino acids are derived from glutamic acid, aspartic acid, L-alanine or L-serine which in turn are synthesized by the amination of their respective keto acid precursors, $\alpha$-oxoglutarate, oxaloacetate, pyruvate and hydroxypyruvate which are either intermediates, or are easily derived from intermediates, of the TCA cycle or glycolysis. Similarly fatty acids are synthesized starting from acetyl CoA via the malonyl CoA pathway and carbohydrates

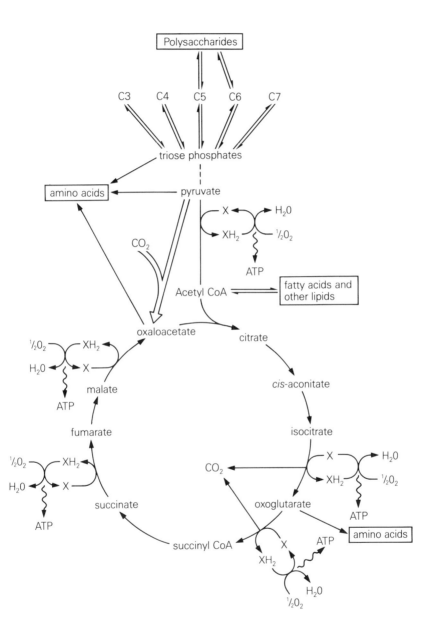

**Fig. 2.1** Diagrammatic outline showing the aerobic catabolism of substrates containing three or more carbons to generate ATP via glycolysis and the TCA cycle and how these substrates can be utilized to generate major biosynthetic precursors (for explanation see text) → catabolic reactions; → biosynthetic reactions; ⇒ anaplerotic reactions

are derived from either malate or oxaloacetate which are decarboxylated to pyruvate or phosphoenolpyruvate. The latter are then further metabolized in a series of reactions, some of which are a reversal of glycolytic reactions and others of which are specialized biosynthetic reactions, to yield six-carbon monosaccharides from which other monosaccharides and polysaccharides can be synthesized. Other cellular requirements, including purine and pyrimidine precursors of nucleotides, are derived more indirectly from TCA cycle and glycolytic intermediates but still start from intermediates of these pathways.

## 2.3   Anaplerotic pathways

It will be apparent from the above that glycolysis and the TCA cycle are extremely important metabolic pathways. Not only do they have a *catabolic*, energy-yielding function but they also provide the major precursors for the synthesis of all types of cellular components (*anabolic function*). This unique dual function of glycolysis and the TCA cycle has been recognized by the introduction of the term *amphibolic* to describe their very important role in cellular metabolism. This role can only be fulfilled if there is a mechanism for the continual regeneration of intermediates of the pathway because, for every two-carbon acetate which enters the TCA cycle, two molecules of carbon dioxide are released. If $\alpha$-oxoglutarate, succinate, malate or oxaloacetate were removed as the precursors for synthesis of cellular components then the cycle would rapidly cease since no oxaloacetate would be a regenerated to accept acetyl CoA. Clearly, therefore, a 'topping-up' or *anaplerotic pathway* is required whereby levels of the TCA cycle intermediates can be replenished. In the vast majority of organisms, using substrates consisting of three or more carbon atoms, this involves a simple carboxylation of a three-carbon compound, usually phosphoenolpyruvate or pyruvate, derived from the substrate itself to yield either malate or oxaloacetate (see Fig. 2.1). Different organisms have different enzymes for achieving this anaplerotic function the most usual being:

$$\text{pyruvate} + CO_2 + ATP + H_2O \xrightleftharpoons{\text{pyruvate carboxylase}} \text{oxaloacetate} + ADP + Pi + 2H^+$$

$$\text{phosphoenolopyruvate} + CO_2 + GDP + Pi \xrightleftharpoons[\text{carboxykinase}]{\text{phosphoenolpyruvate}} \text{oxaloacetate} + GTP$$

As with energy-yielding pathways, organisms growing on many two-carbon substrates or on fatty acids or hydrocarbons with two or more carbons that are essentially oxidized to acetyl CoA have, of necessity, evolved different anaplerotic pathways (see Chapter 4). This is because few organisms have the ability to carboxylate acetate directly to a three-carbon substrate which could itself be anaplerotically carboxylated to either oxaloacetate or malate. The direct reversal of the pyruvate dehydrogenase reaction, which generates acetyl CoA to feed into the TCA cycle, is virtually impossible under normal cellular conditions because of the very large free energy change involved. Some organisms, notably the anaerobic photo-synthetic bacteria, have the ability to convert acetate to pyruvate via a ferredoxin-linked pyruvate synthetase (Stryer, 1981) but since these organisms are of little interest in biotechnology they are not considered further. Organisms growing on one-carbon substrates have an even greater problem of generating the three-carbon and four-carbon compounds required for biosynthesis (see Chapter 3).

## 2.4 Summary

The majority of microorganisms growing on a wide variety of growth substrates oxidize these substrates via glycolysis and the TCA cycle in order to generate reduced nucleotides. In turn these reduced nucleotides are reoxidized via electron transport systems to generate the ATP necessary for cellular maintenance and growth. Biosynthetic pathways leading to the synthesis of carbohydrates, lipids, amino acids and proteins and other cellular components generally begin with intermediates of glycolysis and the TCA cycle. In order to fulfil this dual, amphibolic, role these pathways require a "topping up", anaplerotic, pathway. Normally this involves carboxylation of a three-carbon to a four-carbon compound. Organisms growing on one- and two-carbon compounds, however, require different metabolic pathways which are discussed in subsequent chapters.

# Chapter 3

## *Aerobic microbial growth on one-carbon substrates*

### 3.1 Definition, scope and rationale

The definition of a one-carbon substrate may appear self-obvious. However, in terms of metabolism it includes not only compounds containing only one carbon atom such as carbon dioxide, carbon monoxide, formate, methanol, methane and methylamine, but also compounds with more than one carbon atom, but which have no carbon–carbon bonds such as:

$$
\begin{array}{ccc}
\underset{\text{CH}_3}{\overset{\text{CH}_3}{\diagdown}}\!\text{NH} &
\underset{\text{CH}_3}{\overset{\displaystyle \overset{\text{CH}_3}{\diagdown}}{\diagup}}\!\text{N} &
\underset{\text{CH}_3}{\overset{\text{CH}_3}{\diagdown}}\!\text{S} \\[2ex]
\text{dimethylamine} & \text{trimethylamine} & \text{dimethylsulphide}
\end{array}
$$

In this chapter the energy-yielding pathways found in organisms capable of aerobic utilization of one-carbon substrates will initially be discussed followed by an examination of the biosynthetic routes whereby the substrates are converted to the three-carbon and four-carbon biosynthetic precursors required for growth. A selective treatment is being adopted with a bias towards those micro-organisms, substrates and pathways that are of biotechnological interest. In particular, the utilization of methane, methanol and carbon monoxide is stressed, while utilization of carbon dioxide is given more selective coverage. This is not meant to imply that aerobic organisms capable of utilizing carbon dioxide as their growth substrates (autotrophs) are not important in biotechnology. Indeed plants, algae and cyanobacteria which utilize light energy to generate ATP and reducing power in order to fix carbon dioxide (photoautotrophs) already feature prominently in biotechnological developments. However, the processes of photosynthetic electron transport and photophosphorylation and the subsequent pathway of autotrophic carbon dioxide fixation are well documented in standard biochemistry texts (Lehninger, 1982). Conversely the mechanisms of ATP generation in organisms capable of growing aerobically on methane and methanol (*methylotrophs*) and carbon monoxide (*carboxydotrophs*) are less well documented. Likewise pathways for conversion of one-carbon substrates, other than carbon dioxide, to the three-carbon and four-carbon precursors required for biosynthesis are rarely discussed in general texts.

There is, however, major interest in these organisms and their exploitation. For instance, methylotrophs are already being utilized in biomass production (for example, the growth of *Methylophilus methylotrophus* upon methanol in the, so-called, ICI Prutreen process for biomass production). Additionally, enzymes which have been shown to play key roles in the oxidation of these substrates, especially methane mono-oxygenase, are being exploited in chemical conversions (*bioconversions*) and in *biosensors*. Interest is currently being demonstrated in carboxydotrophs with respect to their possible role in biofilters to remove unwanted carbon monoxide, in bifuel cells aimed at the conversion of chemical energy released during carbon monoxide oxidation to electricity, and in biosensors for detection and quantification of carbon monoxide in various environments including mines, underground car parks and tunnels (Colby, Williams and Turner, 1985). It is clearly important that a full knowledge of the biochemistry of all these organisms is available to underpin studies on their application in biotechnology.

## 3.2   The compounds

Methane ($CH_4$) is the most abundant organic one-carbon compound. It is generated in anaerobic environments such as bottom muds, paddy fields and the ruminant stomach as a result of the terminal stages in the anaerobic breakdown of organic substrates. The process of methane production by *methanogenic bacteria*, which is reviewed in the appendix to this chapter, plays a significant part in the carbon cycle with over 50% of the annual carbon imput into aquatic environments being regenerated as methane (Hanson, 1980). *Methanogenesis* results in the biogenic generation of some 800 million tonnes of methane per annum (Higgins *et al.*, 1981). Methanol ($CH_3OH$) is also generated in significant quantities by anaerobic bacteria, but not to anything like the same level as methane. It is, however, produced in the atmosphere from methane in photo-oxidative processes.

Of the other one-carbon substrates used by aerobic organisms, only carbon monoxide and dimethyl sulphide are produced in significant quantities by natural degradative processes. Large quantities of carbon monoxide are also produced and released into the atmosphere from car exhausts and from blast furnaces. Other one-carbon compounds are generated as byproducts of various industrial processes including formaldehyde (HCHO), formate (HCOOH), methylamine ($CH_3NH_2$), dimethylamine (($CH_3)_2NH$) and trimethylamine (($CH_3)_3N$) from the tanning industry, formate from rubber processing, cyanide (CN) from the electroplating and metal extraction industries and dimethyl sulphide (($CH_3)_2S$), dimethylsulphoxide (($CH_3)_2SO$) and dimethylsulphone (($CH_3)_2SO_2$) from woodpulp processing. In each case the byproducts present a major effluent disposal problem. A knowledge of organisms which can utilize these substrates will, therefore, be of benefit in developing biosensors to detect and quantify them and in the development of economic means of degrading them to less toxic effluents.

### 3.3   Methylotrophs

Prior to the late 1960s relatively few organisms other than autotrophs had been isolated and described that were capable of growing aerobically on one-carbon substrates. Now, however, because of the interest in these organisms, a very large number of species have been isolated and characterized. Those capable of growing on organic one-carbon substrates (in other words, all those substances listed in the previous paragraph except cyanide and carbon monoxide) are known as methylotrophs, while those capable of growing on carbon monoxide are known as carboxydotrophs and those capable of growing on cyanide have been referred to as *cyanotrophs*.

Bacterial methylotrophs have been divided into two types. *Type I organisms* are obligate species which can only grow on methane and/or methanol. They have evenly distributed, disc-shaped membranous inclusions which are particularly well developed when growing on methane, use the ribulose phosphate pathway (RMP) for assimilation of the substrate (see Section 3.10) and have an incomplete TCA cycle. In contrast *type II bacteria* are frequently facultative in that they can utilize not only the full range of one-carbon substrates (usually with the exception of methane), but also a wide range of other substrates with more than one carbon. Their membrane systems are different; being paired and located at the periphery of the cell, they utilize a different assimilatory pathway, the L-serine pathway (see Section 3.11), and have a complete TCA cycle. These marked differences between type I and type II bacterial methylotrophs probably indicate two quite different evolutionary pathways that have converged to result in the utilization of similar substrates.

Type I and type II bacteria are further subdivided into *subgroups A and B,* and into several different genera according to such taxonomic criteria as guanine and cytosine content of the DNA, optimal growth temperature and morphology. Type I organisms fall into three genera: *Methylomonas, Methylobacter* and *Methylococcus. Methylococcus capsulatus* has been the subject of considerable controversy because it has recently been shown to exhibit features of type II as well as type I species. In particular, it is not obligate and will utilize other one-carbon substrates in an autotrophic mode (see Section 3.9). This illustrates the difficulties of classifying a relatively poorly characterized group of organisms at a time when new examples are regularly being isolated. Type II organisms include the obligate genera *Methylosinus* and *Methylocystis*, and amongst the many facultative organisms are species of *Methylobacterium, Methylophilus, Thiobacillus, Hyphomicrobium, Paracoccus*, and numerous pseudomonads.

In addition to bacteria, several yeasts have been isolated from enrichment culture that are capable of growing on methane and methanol including species of *Rhodotorula, Candida, Kloeckera, Hanensula* and *Sporobolomyces*. To date rather less is known about membrane structures and general aspects of metabolism in yeasts than in bacterial methylotrophs.

The membrane systems of bacterial methylotrophs have aroused considerable interest since the extent of their development appears to vary according to growth conditions. They seem to be particularly well developed when growing on methane, but become less developed and less ordered as cultures age and move into the stationary phase of growth. This characteristic is also apparent when other substrates are utilized by facultative species even though the membranes are less well developed in the first place. Similar variations occur in response to oxygen levels and temperature, low oxygen levels and low temperature resulting in increased membrane development. Convincing explanations for these variations have not so far been forthcoming.

## 3.4 Oxidative pathways for ATP generation in bacteria growing on methane, methanol, formaldehyde and formate

There is no point of entry into the TCA cycle for the oxidation of one-carbon substrates (see Section 2.1). Thus a different oxidative pathway, aimed at generating reduced electron acceptors which can feed into electron transport systems to generate ATP, must be used by organisms growing on such substrates. Early studies, carried out largely by Quayle and his coworkers (see Quayle, 1980) using various metabolic inhibitors in organisms growing upon methane, demonstrated that methanol accumulated in the presence of iodoacetate while formaldehyde accumulated in the presence of bisulphite and formate accumulated in the growth medium of resting cells. This led to the proposal of the oxidative pathway illustrated in Fig. 3.1 in which carbon dioxide is the end product. The overall stoichiometry can be summarized as:

$$CH_4 + 2O_2 \rightarrow CO_2 + 2H_2O.$$

This linear pathway has been shown, in subsequent studies, to account for the oxidation not only of methane, but also methanol, formaldehyde and formate as energy substrates, the substrate feeding into the pathway at the appropriate point. Each reaction in this pathway will now be considered in more detail.

### 3.4.1 *Methane mono-oxygenase*
Methane mono-oxygenase proved the most difficult and most interesting reaction of the pathway to characterize. Several possibilities for the oxidation of methane to methanol were initially considered including:

(a)  $CH_4 + O_2 + XH_2 \xrightleftharpoons{\text{methane mono-oxygenase}} CH_3OH + H_2O + X$

(b)  $2CH_4 + O_2 \xrightleftharpoons{\text{methane dioxygenase}} 2CH_3OH$

(c)  $CH_4 + H_2O + X \xrightleftharpoons{\text{methane oxidoreductase}} CH_3OH + XH_2.$

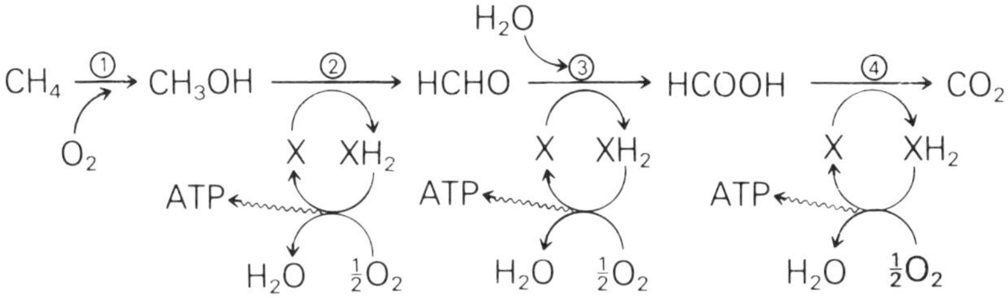

**Fig. 3.1** Oxidative pathway for ATP generation in methylotrophs. X = electron acceptor; $XH_2$ = reduced electron acceptor; (1) methane mono-oxygenase; (2) methanol dehydrogenase; (3) formaldehyde dehydrogenase; (4) formate dehydrogenase

The use of $^{18}O_2$ and $H_2{}^{18}O$ demonstrated that the oxygen in methanol was derived from $O_2$, not $H_2O$, thereby ruling out alternative (c). It was then argued that reaction (b) was more likely than (a) because the available energy for ATP generation would be less with (a) since a reduced nucleotide ($XH_2$) was consumed in the reaction thereby reducing net $XH_2$ production from the pathway and in turn reducing potential production of ATP. Consequently, it was argued that if (a) was used it might be expected that growth yields on methane should be lower than those on methanol whereas, in fact, growth yields were higher in organisms grown on methane. Nontheless, reaction (b) was regarded by many workers as thermodynamically unlikely. The problem of which reaction was involved awaited the development of an *in vitro* system capable of converting methane to methanol. Such a system was eventually developed for *Methylococcus capsulatus*. It was quickly demonstrated that this preparation contained methane mono-oxygenase activity (a).

Further studies on this enzyme system have continued to be controversial and it now appears there are two forms of the enzyme, one membrane-bound and one soluble. The soluble system was demonstrated first in *Methylococcus capsulatus* and the particulate system in *Methylosinus trichosporium*, but it now appears that both enzymes may be present in both species dependent on growth conditions and in particular the nature of the limiting growth factor ($O_2$, $CH_4$ or N). More recently Stanley *et al.* (1983) have shown in *Methylococcus capsulatus*, previously assumed to have only the soluble enzyme, that the availability of copper appears to influence the cellular location and type of methane mono-oxygenase. In cells grown at low copper levels the enzyme is soluble, but it becomes particulate when grown at high copper levels. Moreover, the protein composition of the enzyme changes.

The soluble enzyme of *Methylococcus capsulatus* consists of three proteins, A, B and C. *Protein A* is a non-haem iron and sulphur-containing protein, while *protein B* contains no known prosthetic group and *protein C* is a flavoprotein also containing iron and sulphur. The postulated reaction sequence for the oxidation of methane for this enzyme is:

where FPC is flavoprotein C.

The particulate enzyme is less well characterized, but in *Methylosinus trichosporium* it is also composed of three proteins. Interestingly the purified enzyme does not use NAD(P)H as the electron donor. Instead, one of the three component proteins, reduced cytochrome *c*, acts in this capacity.

Clearly further studies are required on both types of methane mono-oxygenase in order to fully characterize the mechanism of the oxidation of methane to methanol and to resolve the question of cellular location and type of enzyme. In particular, it will be interesting to establish whether the ability to switch from one type to the other is a widespread phenomena, to clarify the significance of such changes and to examine the regulatory mechanism involved. In addition studies are required on the enzyme isolated from a broader range of methylotrophs since, in the yeast *Candida*, cytochrome $P_{450}$ appears to substitute for protein B in the scheme outlined above, whilst in *Pseudomonas putida*, a different protein *rubredoxin* appears to replace *both* proteins A and B.

There is little doubt, however, that the necessary clarification will be forthcoming because of the intense interest in the exploitation of methane mono-oxygenase activity in a wide variety of biotechnological processes (Dalton, 1980; Higgins, Best and Hammond, 1980; Large, 1983). This interest stems from the important and somewhat unusual observation that the enzyme has a very wide range of substrates which it will utilize including long and short-chain alkanes, alkenes and aromatic and heterocyclic substances such as cyclohexane, benzene, toluene and pyridine. It is predicted that the enzyme could be used in the catalysis of reactions which, by normal chemical processes, are very difficult and/or give low yields. Hence the enzyme could be of enormous value in the chemical industry. Moreover, *recalcitrant molecules*, which are not readily biodegradable but persist in the environment causing pollution problems, may be degraded to more biodegradable products by this enzyme thereby alleviating some of their non-desirable environmental effects.

The wide substrate specificity of methane mono-oxygenase allows obligate methylotrophs possessing this enzyme the ability to oxidize and, presumably, to benefit at least in terms of ATP production, from other substrates which themselves cannot support growth. This has been discussed by Higgins, Best and Hammond (1980) who have argued against use of the term *fortuitous metabolism*, which had been applied to this phenonemon, on the grounds that some benefit must accrue to the organism. The term *co-metabolism* is considered preferable.

### 3.4.2 *Methanol dehydrogenase*

Methanol dehydrogenase in bacteria is less variable than methane mono-oxygenase differing from species to species only in substrate affinity and molecular weight. In all cases the enzyme will oxidize primary alcohols with a chain length up to at least $C_{11}$. In some species the enzyme also exhibits formaldehyde dehydrogenase activity. A characteristic of methanol dehydrogenases is that they all contain a quinone-based prosthetic group called either *methoxatin* or *pyrrolo-quinoline quinone* (PQQ) and hence they are referred to as one of only a small number of *quinoproteins* which have been isolated to date. This prosthetic group appears to act as the primary electron acceptor in the oxidation of methanol to formaldehyde and in turn it transfers electrons to cytochrome *c*. The latter is reoxidized by a terminal oxidase which is either cytochrome $a/a_3$ or cytochrome *o*.

### 3.4.3 *Formaldehyde dehydrogenase*

Formaldehyde dehydrogenase is a common enzyme found in most organisms since formaldehyde is a normal metabolite involved in many cellular reactions. As indicated above, methanol dehydrogenase may exhibit formaldehyde dehydrogenase activity. In most species, however, a separate $NAD^+$-dependent formaldehyde dehydrogenase has been characterized. The subsequent electron transport system is not well known though it is believed cytochrome *c* is involved and again either cytochromes $a/a_3$ or *o* act as the terminal oxidase.

Some type I bacteria which use the RMP pathway of formaldehyde assimilation (see Section 3.10) have very low activities of formaldehyde dehydrogenase and of the next enzyme in the oxidative sequences, formate dehydrogenase. It has been demonstrated that these organisms may oxidize formaldehyde via the RMP pathway and 6-phosphoglycerate dehydrogenase in a specialized oxidative version of the pathway. Discussion of this alternative oxidative route is more appropriate when the RMP pathway has been outlined (see Section 3.10).

### 3.4.4  *Formate dehydrogenase*

Formate dehydrogenase is also an enzyme of widespread occurrence. In most methylotrophs, it has been shown to be $NAD^+$-dependent linking to cytochrome $c$ and then to a terminal oxidase.

It will be clear from the above account that there is considerable variation in individual bacteria with respect to the precise mechanisms of the reactions responsible for the oxidation of methane to carbon dioxide. In particular, our knowledge of the electron acceptors/donors, of the electron transport chains involved and their coupling to oxidative phosphorylation to yield ATP is far from clear. Undoubtedly ATP yields are high with each of the three dehydrogenase catalysed steps releasing sufficient energy for the generation of 3 ATP making a total of at least 9 ATP per molecule of methanol oxidized to $CO_2$. This contrasts with the generation of 6 ATP for each of the carbon atoms in glucose when it is completely oxidized to $CO_2$ via glycolysis and the TCA cycle. Thus, the energy yield from methane and methanol oxidation is extremely good and explains the high growth yields for most bacteria growing on these substrates compared to more oxidized substrates.

## 3.5   Oxidative pathways for ATP generation in yeasts growing on methanol

Methylotrophic yeasts grow only very slowly on methane and most work has been done on their ability to oxidize methanol, though it is possible these organisms can also utilize other one-carbon compounds. Why, it may be asked, is the oxidative pathway in yeasts being dealt with separately from that in bacteria? The answer is that the details of the pathway are very different with the first reaction, the conversion of methanol to formaldehyde, being catalysed not by a dehydrogenase but by *methanol* (*alcohol*) *oxidase*. This enzyme, which is located in peroxisomes, is a FAD-linked flavoprotein which consists of eight identical subunits and which oxidizes a variety of alcohols besides methanol. The overall reaction for methanol oxidation is:

$$CH_3OH + O_2 \rightarrow HCHO + H_2O_2.$$

Because of its toxicity the hydrogen peroxide released is immediately removed by peroxisomal catalase and therefore the overall reaction is effectively irreversible. It will be noted that this reaction does not involve the reduction of an electron acceptor which could be reoxidized via electron transport mechanism to yield ATP. Thus, ATP yields are lower in yeasts than in bacteria and this probably contributes to the slower growth rates and lower growth yields characteristic of yeasts using methanol (Egli and Lindley, 1984).

Formaldehyde, the product of the methanol oxidase reaction, is oxidized, as in bacteria, to formate. Again, the reaction involved is different from that in bacteria. The formaldehyde dehydrogenase enzyme in yeasts is both $NAD^+$ and glutathione

(GSH)-dependent and the reaction proceeds via the initial formation of S-hydroxy-methylglutathione and results in the synthesis of formylglutathione:

$$HCHO + GSH + NAD^+ \rightarrow GS \sim OCH + NADH + H^+.$$

In *Candida boidinii* the thiol-linked ester is then hydrolysed by an *esterase* to regenerate GSH and release formate:

$$GS \sim OCH + H_2O \rightarrow GSH + HCOOH.$$

Formate dehydrogenase finally oxidizes the formate to carbon dioxide in a reaction which appears to be identical to that in bacteria. Both formaldehyde dehydrogenase and formate dehydrogenase activity in yeast has been shown to be located exclusively in the cytoplasm. Thus, formaldehyde produced as a result of methanol oxidase activity in the peroxisome must leave the organelle for subsequent oxidation and the NADH generated by the two dehydrogenases must enter mitochondria to be reoxidized via electron transport systems. Current evidence suggests that the electron transport system involved uses the NADH oxidase system in mitochondria which generates not three, but only two ATP per pair of electrons transported. Again this would account for the lower growth yields of yeasts growing on methanol compared to bacteria.

## 3.6 Oxidative pathways for ATP generation in bacteria growing on other one-carbon substrates (excluding carbon monoxide and cyanide)

Several studies have examined the growth, particularly of pseudomonads, upon the methylated amines methylamine, dimethylamine and trimethylamine. It has been established in each case is that the methyl group is converted to formaldehyde which is then further oxidized, as in methylotrophic bacteria growing on other one-carbon substrates, via formate to carbon dioxide (see Section 3.4). However, there seems to be a remarkable variety of ways in which the conversion to formaldehyde can be achieved.

For instance, with methylamine, at least three different pathways are known (Large, 1981):

(i) $CH_3NH_3^+ + O_2 + H_2O \xrightarrow{\text{amine oxidase}} HCHO + H_2O + NH_4^+$

(ii) $CH_3NH_3^+ + H_2O + X \xrightarrow[\text{dehydrogenase}]{\text{methylamine}} HCHO + NH_4^+ + XH_2$

(iii) $CH_3NH_3^+ + glutamate \xrightarrow[\text{synthetase}]{\text{methylglutamate}} methylglutamate + NH_4^+$

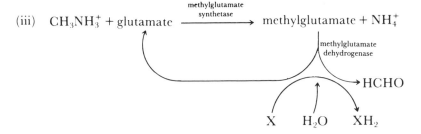

With dimethylamine, two alternative routes have been established:

(i)   $(CH_3)_2NH_2^+ + O_2 + NAD(P)H + H^+ \xrightarrow[\text{mono-oxygenase}]{\text{dimethylamine}}$

$$CH_3NH_3^+ + HCHO + NAD(P)^+ + H_2O$$

(ii)   $(CH_3)_2NH_2^+ + X + H_2O \xrightarrow[\text{dehydrogenase}]{\text{dimethylamine}} CH_3NH_3^+ + HCHO + XH_2.$

The mono-oxygenase reaction is the most common. This enzyme is dependent on oxygen while the alternative dehydrogenase can proceed anaerobically provided a suitable system of reoxidizing the reduced electron acceptor is available (for example, $NO_3^-$). The formaldehyde generated by both these reactions is oxidized as previously, while the methylamine is oxidized to formaldehyde prior to further oxidation.

Similarly with trimethylamine, the methyl groups are converted, one at a time, to formaldehyde. Again the initial oxidation of one methyl group to formaldehyde can proceed in one of two different ways using a mono-oxygenase or dehydrogenase. The mono-oxygenase system, however, at least in *Pseudomonas aminodovorans*, proceeds in a different two-step reaction in which trimethylamine-N-oxide $(CH_3)_3NO$ is formed and then converted to dimethylamine and formaldehyde. Organisms using this route can also utilize trimethylamine-N-oxide as their energy and carbon source.

Only relatively few, strictly aerobic, bacteria can oxidize tetramethylammonium salts. The key reaction, oxidation of one methyl group to formaldehyde, is catalysed by a mono-oxygenase, and yields trimethylamine as the other product. The enzyme uses $NAD(P)H$ as the electron donor.

The variety of mechanisms for oxidizing methylated amines is somewhat surprising, but it should be noted that in each case a mono-oxygenase with a broad spectrum of substrates is usually the main mechanism. On the other hand it is significant that organisms using the dehydrogenase mechanism usually grow better on these substrates than organisms employing the mono-oxygenase system. This is undoubtedly related to the fact that mono-oxygenases *consume* reduced electron acceptors, thereby reducing the level of reduced acceptors available to enter electron transport and hence be used in ATP generation. Conversely, the dehydrogenase system *produces* reduced electron acceptors thereby allowing more ATP to be produced during their reoxidation.

The oxidation of methylated sulphur compounds has received less attention than other one-carbon substrates. *Pseudomenas* MS grows on trimethylsulphonium salts, in a way quite different from that for growth on trimethylamine, by transferring one methyl group to the one-carbon carrier compound tetrahydrofolate (THF). This subsequently undergoes a series of conversions to form formyl THF which is then converted to formate along with the regeneration of the THF acceptor. The dimethylsulphide, which is the other product of the initial reaction, appears not to be further oxidized by this organism. Recently a species of *Hyphomicrobium* has been isolated which can utilize dimethylsulphoxide, not only as sole energy and carbon source, but also as sole sulphur source. Clearly, in this case, the substrate is being completely oxidized not only for ATP generation, but also to release sulphur, probably as sulphate, which is essential for growth. The enzyme reactions involved have not yet been characterized. However, mechanisms of utilization of methylated sulphur compounds are arousing considerable interest, because of their potential role in desulphurization of coal, etc.

### 3.7   Carboxydotrophs and cyanotrophs

It may come as a surprise that two very toxic substances, carbon monoxide and cyanide, can be used as the sole carbon substrate for growth by some micro-organisms called *carboxydotrophs* and *cyanotrophs* respectively. Both these substances are potent inhibitors of cytochrome $a/a_3$ (cytochrome oxidase) and prevent this terminal electron transport protein being reoxidized by molecular oxygen. This has a catastrophic effect on respiration, since reduced nucleotides cannot be reoxidized aerobically and ATP generation is limited. Thus, both carbon monoxide and cyanide are normally extremely toxic to living organisms. Those organisms which can grow on these substrates must therefore have an electron transport mechanism independent of normal cytochrome $a/a_3$ activity.

To date, relatively little work has been published on growth using cyanide as substrate other than to demonstrate that several pseudomonads and some actinomycetes will utilize it as both sole carbon and nitrogen source. However, much more work has been done on the carboxydotrophs (Meyer and Schlegel, 1983). These appear to be a taxonomically diverse group of bacteria most of which also seem to be able to grow autotrophically as hydrogen bacteria (that is, $H_2 + CO_2$). Among the organisms shown to be able to utilize carbon monoxide are *Pseudomonas gazotropha*, which also grows well on methanol, *Pseudomonas carboxydovorans* which will also grow on acetate, pyruvate and some other TCA cycle intermediates, *Bacillus schlegelii* and species of *Azotobacter*. These organisms have been shown to be of ubiquitous distribution in soils, freshwater muds and in coal mines. They can tolerate remarkably high levels of carbon monoxide (up to 90%).

The oxidation of carbon monoxide to carbon dioxide is catalysed by the enzyme *carbon monoxide oxidase*:

$$CO + H_2O + X \rightarrow CO_2 + XH_2.$$

The $XH_2$ generated is reoxidized via an electron transport system to yield the ATP necessary for growth. The assimilatory pathway is the ribulose bisphosphate pathway (RBP or Calvin cycle) of carbon dioxide fixation (see Section 3.9). Growth on carbon monoxide therefore involves a specialized form of autotrophy in which the substrate is first completely oxidized to carbon dioxide, which is then fixed by the conventional autotrophic route using energy and reducing power generated by the initial oxidation. In such a situation it is clear that the efficiency of growth is going to be very low (generation time of 12–42 h). Consequently biotechnological interest in using carbon monoxide as a growth substrate for bacteria does not involve its use as a potential source of biomass production.

The enzyme carbon monoxide oxidase is a complex protein being a unique selenium activated, molybdo–iron–sulphur FAD-dependent flavoprotein. The molybdenum group is associated with a novel pterin. The subsequent reoxidation of this protein, whose mechanism of action is not yet understood (Meyer and Schlegel, 1983), involves a carbon monoxide-insensitive electron transport system involving cytochrome $b_{561}$ and in which the terminal electron acceptor is cytochrome $o$ (sometimes referred to as cytochrome $b_{563}$).

It should be noted that certain methylotrophs can also oxidize carbon monoxide, though inefficiently, by using the broad-spectrum activity of methane monooxygenase. These organisms are not, however, able to grow on carbon monoxide as the sole carbon source. In contrast, several anaerobes can grow on carbon monoxide including *methanogens* (see the appendix to this chapter), *homoacetogens*, such as *Clostridum pasteurianum*, and some photosynthetic bacteria including *Rhodopseudomonas* sp. These species use a *carbon monoxide dehydrogenase*, which catalyses the same reaction as carbon monoxide oxidase. This enzyme has quite different characteristics from carbon monoxide oxidase and is extremely oxygen-sensitive.

### 3.8    The assimilatory routes of organisms utilizing one-carbon substrates

As outlined in Section 2.2, all organisms require to generate essential three-carbon and four-carbon compounds, mostly intermediates of the glycolytic pathway and the TCA cycle, which are the precursors for biosynthesis of cellular components. This is generally achieved by the anaplerotic carboxylation of a three-carbon compound to yield oxaloacetate or malate (see Section 2.3). However, in organisms growing on one-carbon compounds three-carbon compounds are not generated in the oxidative reactions. Thus an assimilatory mechanism of generating three-carbon and four-carbon intermediates for biosynthesis is required. A wide variety of such pathways have been elucidated in different organisms to achieve this objective (dealt with in Sections 3.10 to 3.13). For more detailed accounts the reader should consult recent reviews (such as Quayle, 1980; Higgins *et al.*, 1981; Large, 1983).

### 3.9    The ribulose bisphosphate (RBP) pathway for carbon dioxide assimilation

All autotrophic organisms including plants, utilize the ribulose bisphosphate (RBP) pathway for reducing carbon dioxide. This pathway was originally established by Calvin and his co-workers and is often referred to as the Calvin cycle (Lehninger, 1982). In this pathway, which is given in essential outline in Fig. 3.2, ribulose, 1-5-bisphosphate (RBP), acts as an acceptor for carbon dioxide and is carboxylated to form a transient six-carbon monosaccharide phosphate which is very unstable and instantly splits to two molecules of the three-carbon compound phosphoglycerate. This is subsequently phosphorylated to bisphosphoglycerate and then reduced to glyceraldehyde-3-phosphate (G-3-P). As shown in Fig. 3.2, for every three molecules of carbon dioxide fixed three molecules of the acceptor (RBP) are required and six molecules of G-3-P are produced. However, to keep the pathway operating the three, five-carbon, acceptor molecules must be regenerated. This involves the rearrangement of 15 carbons in five molecules of G-3-P via a series of monosaccharide phosphates and leaves a net production of one molecule of G-3-P. This can readily be converted via glycolysis to phosphoenolpyruvate or pyruvic acid and hence can serve as the source of three-carbon and four-carbon precursors required for biosynthesis exactly as in any other organism (see Sections 2.2, 2.3 and Fig. 2.1).

Full details of the pathway are found in most biochemistry text books so are not given here. It should be noted that the two crucial enzymes unique to this pathway are the initial, carboxylating enzyme *ribulose bisphosphate carboxylase* (carboxy dismutase) and *ribulose-5-phosphate kinase* which is responsible for the last step in the regeneration of RBP. The overall stoichiometry of this pathway is usually written as:

$$3CO_2 + 6NAD(P) + 6H^+ + 9ATP \rightarrow G\text{-}3\text{-}P + 6NAD(P)^+ + 9ADP + 8Pi.$$

Other pathways of one-carbon assimilation have end-products other than G-3-P. In order to compare the stoichiometry of these various pathways it is essential to use a common end product, namely pyruvate. If the G-3-P produced in the RBP pathway is subsequently oxidized to pyruvate via the glycolytic pathway, one $NADH_2$ and 2ATP

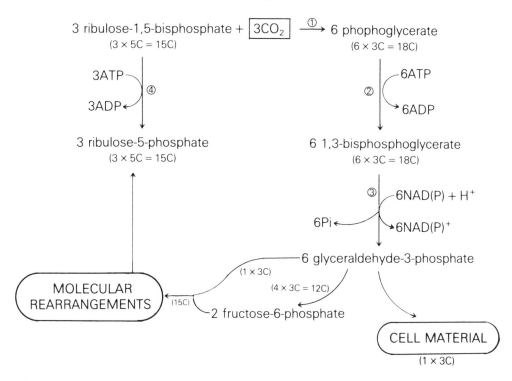

**Fig. 3.2** Outline of the ribulose bisphosphate (RBP) pathway for autotrophic carbon dioxide fixation. ① ribulose-1,5-bisphosphate carboxylase; ② phosphoglycerate kinase; ③ glyceraldehyde-3-phosphate dehydrogenase; ④ ribulose-5-phosphate kinase

are produced. Thus the stoichiometry of the RBP pathway for pyruvate production becomes:

$$3CO_2 + 5NAD(P)H + 5H^+ + 7ATP \rightarrow pyruvate + 5NAD(P)^+ + 7ADP + 7Pi.$$

Clearly therefore the RBP pathway is very energy-demanding consuming five reducing equivalents and 7 ATP in order to produce, from three molecules of carbon dioxide, one molecule of pyruvate. As will be seen, this stoichiometry is unfavourable when compared to other pathways of one-carbon assimilation (see Sections 3.10, 3.11, 3.12).

The RBP pathway is not exclusively found in strictly autotrophic organisms. For instance, as suggested in Section 3.7, it is the route by means of which aerobic carboxydotrophs assimilate the carbon dioxide which results from the oxidation of carbon monoxide as their sole carbon substrate. In addition certain methylotrophs growing on formate, including *Pseudomonas oxalaticus*, *Methylococcus capsulatus* and *Paracoccus denitrificans* (the latter also when utilizing methanol) have been shown to oxidize the substrate to carbon dioxide and then to fix the carbon dioxide released via the RBP pathway. In view of the energy demands of this pathway, and since only one $NAD^+$ will be reduced for every formate molecule oxidized via formate dehydrogenase to carbon dioxide, it is not surprising that growth yield is poor. Presumably this is offset by the increased range of substrates these organisms are able to utilize.

## 3.10   The ribulose monophosphate (RMP) pathway of formaldehyde assimilation in type I bacteria

It was originally anticipated that all methylotrophs would oxidize their substrates to carbon dioxide and then assimilate this autotrophically via the RBP pathway. However, in type I bacteria, such as *Methylomonas methanica*, the pattern of $^{14}CO_2$ fixation and $^{14}C$-methane oxidation were shown not to be identical. Moreover, the crucial enzyme of the RBP pathway ribulose–bisphosphate carboxylase, was not present in these organisms when utilizing methane or methanol. The first labelled product of $^{14}C$-methane assimilation was eventually identified as the six-carbon monosaccharide phosphate *hexulose-6-phosphate* (H-6-P) and an enzyme system capable of forming H-6-P *in vitro*, using ribulose monophosphate as the acceptor of formaldehyde was eventually isolated. This enzyme, *hexulose-phosphate synthetase*, is present in all type I bacteria growing on methane, methanol, formaldehyde and substrates which are oxidized to carbon dioxide via formaldehyde (for example, methylamine etc.) and is unique to this pathway. H-6-P is further metabolized in a sequence of reactions (see Fig. 3.3) bearing considerable similarity to the RBP pathway. Initially three molecules of H-6-P are converted to their isomer fructose-6-phosphate (F-6-P) by another unique enzyme *hexulose–phosphate isomerase*. One F-6-P molecule is phosphorylated to fructose-1-6-bisphosphate prior to its cleavage to form two molecules of glyceraldehyde-3-phosphate (G-3-P). One of these and the two other molecules of F-6-P join in a series of rearrangements to regenerate the three acceptor molecules of ribulose monophosphate and thereby enable the pathway to continue. Thus a net production of one triose molecule results which fulfils the demand for three-carbon and four-carbon compounds required as biosynthetic precursors.

There are subtle differences in the detail of the overall reaction sequence of the RMP pathway in various type I bacteria. These involve the mechanism of conversion of F-6-P to two, three-carbon, compounds and the molecular rearrangements involved in the regeneration of RMP. The variants for F-6-P metabolism are the:

(i) *glycolytic version* which generates two triose phosphate molecules via fructose-1,6-bisphosphate and the
(ii) *Etner–Doudoroff version* which converts F-6-P to triose phosphate ($1 \times 3C$) and pyruvate ($1 \times 3C$).

The variants for the regeneration of the RMP are the: (a) *transaldolase* and (b) *sedoheptulose phosphatase* versions.

The striking similarity of the RBP and RMP pathways should not be allowed to mask the very significant differences in their stoichiometry. As will be seen from Fig. 3.3, to generate one molecule of G-3-P via the *glycolysis, transaldolase* version of the pathway the overall stoichiometry is:

$$3HCHO + ATP \rightarrow G\text{-}3\text{-}P + ADP.$$

If this is taken to the comparative end-point chosen earlier (pyruvate) the stoichiometry becomes:

$$3HCHO + ADP + Pi + NAD(P)^+ \rightarrow pyruvate + ATP + NAD(P)H + H^+.$$

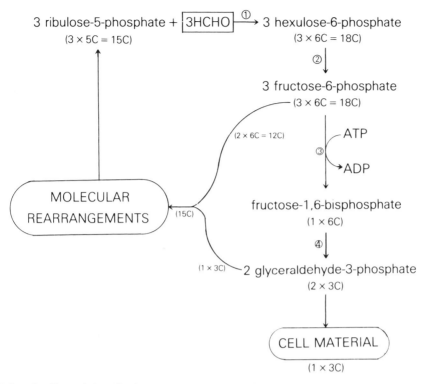

**Fig. 3.3** Outline of the ribulose monophosphate (RMP) pathway for the assimilation of formaldehyde in type I bacteria. ① hexulose–phosphate synthetase; ② hexulose–phosphate isomerase; ③ fructose phosphate kinase (phosphofruktokinase); ④ fructose–bisphosphate aldolase. NB: The pathway illustrated is the glycolysis version. In the Etner–Duodoroff version, fructose-6-phosphate is converted by hexose phosphate isomerase to glucose-6-phosphate dehydrogenase and this is dehydrated to 2-keto-3-deoxy-6-phosphogluconate by 6-phosphogluconate dehydrogenase prior to the reaction catalysed by 2-keto-3-deoxy-6-phosphogluconate aldolase which produces two three-carbon compounds glyceraldehyde-3-phosphate and pyruvate

For the other versions the overall stoichiometry is:

(i) *glycolysis/sedoheptulose phosphatase* and *Etner–Doudoroff/transaldolase*

$$3HCHO + NAD(P)^+ \rightarrow pyruvate + NAD(P)H + H^+$$

(ii) *Etner–Doudoroff/sedoheptulose phosphatase*

$$3HCHO + 3ATP + NAD(P)^+ \rightarrow pyruvate + NAD(P)H + H^+ + 3ADP + 3Pi.$$

Whatever version of the pathway is used it is remarkably efficient when compared to the RBP pathway of carbon dioxide fixation which demands a high imput of both ATP and $NAD(P)H + H^+$ (see Section 3.9). Since the most efficient version is the glycolysis/transaldolase route it is not surprising that organisms exhibiting this route are preferred in industrial processes aimed at biomass production when using methane or methanol as substrates. The main organism which has been exploited to date is *Methylophilus methylotrophus* which, like most other methylotrophs currently being investigated for biomass production, uses the RMP pathway. However, it was chosen before details were known of which version of the RMP pathway it utilized. It now seems it is the Etner–Doudoroff/transaldolase route (Beardsmore, Aperghis and Quayle, 1982) which is slightly less efficient than the glycolysis/transaldolase version. It is conceivable, therefore, that attempts may be made in future to genetically engineer the glycolysis/transaldolase system into this organism. In this context, it is significant that *M. methylotrophus* has already been subject to genetic manipulation to improve its growth on methanol. Ammonia, essential for amino acid synthesis, is incorporated into glutamate via one of two reactions:

(i) $\text{oxoglutarate} + NH_4^+ + NAD(P)H + H^+$

$$\xrightleftharpoons[]{\text{glutamate dehydrogenase}} \text{glutamate} + NADP^+ + H_2O$$

and

(ii) $\text{glutamate} + NH_4^+ + ATP \xrightleftharpoons[]{\text{glutamine synthetase}} \text{glutamine} + ADP + Pi$

followed by

$\text{glutamine} + \text{oxoglutarate} + NAD(P)H + H^+$

$$\xrightleftharpoons[]{\text{glutamate synthetase}} 2 \text{ glutamate} + NADP^+ + H_2O.$$

Many bacteria, including some methylotrophs (for example, *Methylomonas methanica*), possess both systems with the former being operative under conditions of nitrogen excess and the latter under conditions of low nitrogen. This is because glutamine synthetase has a much higher affinity for ammonia than the former. However, *Methylophilus methylotrophus* was found to only contain the glutamine synthetase/glutamate synthetase system. This is a more energy-demanding, consuming ATP which the glutamate dehydrogenase system does not. Thus, Windass *et al.* (1980), isolated a mutant of the organism lacking glutamate synthetase and used a cloning vector to engineer the *E. coli* gene for glutamate dehydrogenase into it. The resulting 'engineered' strain of *M. methylotrophus* has up to a 7% higher yield than the wild type reflecting the significance of conserving just one ATP for each glutamate synthesized.

It is also worthy of mention that *Methylococcus capsulatus*, which has previously been referred to as having characteristics of both type I and type II organisms, contains not only the RMP pathway but is also able to use an active ribulose bisphosphate carboxylase to fix carbon dioxide via the RBP pathway and hence use an autotrophic mode of nutrition. The RMP pathway, as outlined above, is much more energetically favourable than the RBP pathway and the role of the second one-carbon fixing system in this organism requires clarification since it seems unusual that it contains both pathways.

It was mentioned previously (see Section 3.4) that some organisms (for example, *Methylophilus methylotrophus*) which utilize the RMP for the assimilation of one-carbon substrates have low activities of formaldehyde and formate dehydrogenase which are key enzymes in the normal oxidative pathway for energy production and that these organisms may oxidize their substrates via an oxidative RMP pathway. In these organisms particularly active *glucose-6-phosphate* and *6-phosphonogluconate dehydrogenases* have been detected when growing on methanol and methylamine but not when growing on methane. Thus it has been suggested that when utilizing the former substrates a *dissimilatory RMP* is involved (see Fig. 3.4). It should be stressed that, as yet, full confirmatory evidence of this route is awaited and that it is still possible that the bulk of substrate oxidation proceeds via the normal oxidative route, including formaldehyde and formate dehydrogenases even though their activity is low.

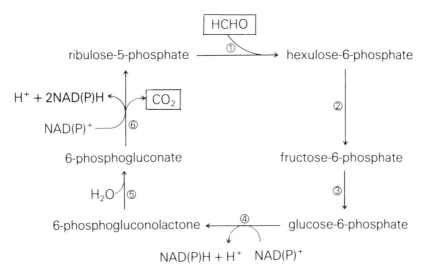

**Fig. 3.4** The dissimilatory ribulose monophosphate pathway which may be used by some type I methylotrophs as the oxidative, energy yielding, route of one-carbon metabolism. ① hexulose–phosphate synthetase; ② hexulose–phosphate isomerase; ③ hexose–phosphate isomerase; ④ glucose-6-phosphate dehydrogenase; ⑤ 6-phospho-gluconolactonase; ⑥ 6-phosphogluconate dehydrogenase

## 3.11 The serine pathway of formaldehyde assimilation in type II bacteria

The serine pathway of formaldehyde assimilation in type II bacteria was first demonstrated in *Pseudomonas* AMI. It is quite different from both the RBP and RMP pathways, especially since only 50% of the assimilated carbon comes directly from formaldehyde with the rest coming from carbon dioxide. However, as will be shown, this carbon dioxide fixation does not involve ribulose bisphosphate carboxylase which is not present in type II organisms growing on one-carbon substrates. The initial products of [14]C-methanol (or [14]C-methylamine) assimilation are one-carbon derivatives of the coenzyme tetrahydrofolic acid (THF) which is a common carrier of one-carbon units in many one-carbon transfer reactions that occur in all living organisms.

As shown in Fig. 3.5 the methylene form of THF acts as the one-carbon donor, the acceptor compound being glycine and the product L-serine. L-serine is then deaminated to its keto-acid equivalent hydroxypyruvate which is reduced by an enzyme regarded as characteristic of the L-serine pathway, *hydroxypyruvate reductase*, to form glycerate. This is then phosphorylated and the resulting phosphoglycerate is further metabolized to form phosphoenolpyruvate and pyruvate via the normal

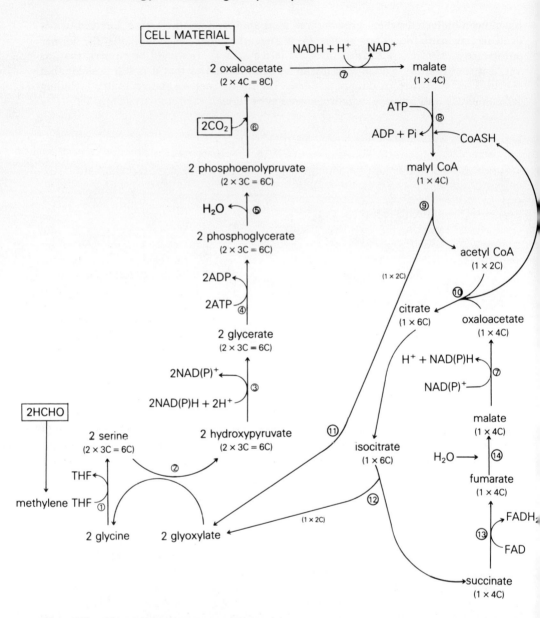

**Fig. 3.5** The isocitrate lyase (positive) version of the serine pathway for the assimilation of formaldehyde in type II bacteria. ① serine hydroxymethyltransferase; ② serine–glyoxylate aminotransferase; ③ hydroxypyruvate reductase; ④ glycerate kinase; ⑤ phosphoenolpyruvate hydratase; ⑥ phosphoenolpyruvate carboxylase; ⑦ malate dehydrogenase; ⑧ malyl–CoA synthetase; ⑨ malyl–CoA lyase; ⑩ citrate synthetase; ⑪ aconitase; ⑫ isocitrate lyase; ⑬ succinic dehydrogenase; ⑭ fumarate reductase

reactions of glycolysis. Either of these compounds may be carboxylated to form oxaloacetate using *PEP carboxylase* (the usual system), *PEP carboxykinase* or *pyruvate carboxylase*.

At this stage the reaction system can be summarized as:

$$2HCHO + 2CO_2 + 2CH_2NH_2COOH \rightarrow 2 \begin{array}{c} CH_2COOH \\ | \\ CH_2COOH \end{array} + NH_3$$

and satisfies, via a very different mechanism from the RBP and RMP pathways, the requirement of organisms growing on one-carbon substrates to generate three-carbon and four-carbon precursors for biosynthesis. However, to achieve a continual flow of carbon through the pathway the two glycine acceptor units must be regenerated from one of the oxaloacetate molecules, leaving a net production of one oxaloacetate available for biosynthesis.

Various mutation studies have shown that glycine regeneration involves the transamination of glyoxylate, but the origin of glyoxylate remained unresolved for some time. Eventually, as with nearly all the pathways involved in growth of different organisms upon one-carbon substrates, two major variants for the regeneration of glyoxylate from oxaloacetate were demonstrated. The main route (see Fig. 3.5), found in *Pseudomonas* MA, *Pseudomonas* MS and *Pseudomonas aminovorans*, is dependent on two further enzymes regarded as characteristic of this version of the L-serine pathway. One of these is *malyl CoA lyase* which splits malyl CoA, derived from oxaloacetate via malate, to two, two-carbon compounds one of which is glyoxylate. The other two carbon unit produced, acetyl CoA, must itself be used to regenerate the second glyoxylate. Thus it condenses with oxaloacetate and is metabolized via TCA cycle reactions to isocitrate at which point the second crucial enzyme, *isocitrate lyase,* splits the six-carbon acid to form glyoxylate and succinate. The latter is metabolised via normal TCA reaction to regenerate the original oxaloacetate acceptor. This pathway is known as the *isocitrate lyase* ($+ve$) version of the L-serine pathway.

Some type II methylotrophs, including *Ps* AMI and *Hyphomicrobium* do not induce isocitrate lyase when growing on one-carbon substrates. Instead they use an *isocitrate lyase* (*negative*) version of the L-serine pathway in which the acetyl CoA which condenses with the five-carbon oxoglutarate to form the seven-carbon acid homo-isocitrate. This is split by *homoisocitrate lyase* to glyoxylate and glutarate, and the glutarate is subsequently metabolized to regenerate the initial oxoglutarate acceptor.

The overall stoichiometry for the production of pyruvate, in both versions of this pathway is identical:

$$2HCHO + 2CO_2 + 2NAD(P)H + 2H^+ + 2ATP \rightarrow pyruvate + 2NAD(P)^+ + 2ADP + 2Pi.$$

This is somewhat less efficient than all four versions of the RMP pathway (see Section 3.10), but is considerably more favourable than the autotrophic RBP pathway (see Section 3.9). It must also be remembered, from the point of view of biomass production, that a far greater range of species can utilize, often facultatively, one-carbon growth substrates via this route than via the RMP pathway.

## 3.12 The xylulose phosphate pathway of formaldehyde assimilation in methylotrophic yeasts

As outlined in Section 3.5, methylotrophic yeasts oxidize methanol via a different route from that found in bacteria. When it comes to the assimilatory pathway, once again

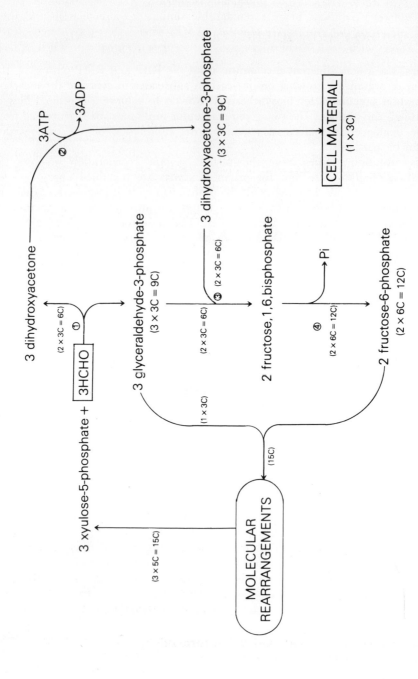

**Fig. 3.6**  Outline of the xylulose monophosphate (XMP) pathway for assimilation of formaldehyde in methylotrophic yeasts. ① dihydroxyacetone synthetase; ② triokinase; ③ fructose-1,6-bisphosphate aldolase; ④ fructose-1,6-bisphosphatase

yeasts are different, though for some time it was believed they used the RMP pathway. However, although the pattern of $^{14}$C-methanol fixation was remarkably similar to that found in type I bacteria, the marker enzymes of the RMP pathway, hexulose phosphate synthetase and hexulosephosphate isomerase, were not found in yeasts. Instead a different pathway, bearing remarkable similarity to these RBP and RMP pathways, known as the *xylulose monophosphate* (*XMP*) *pathway* has been demonstrated (see Fig. 3.6). As with the RMP pathway, formaldehyde is the precursor of cell material and it is transferred to the five-carbon acceptor, xylulose-5-phosphate. The products of this reaction are two three-carbon compounds glyceraldehyde-3-phosphate and dihydroxyacetone phosphate. The crucial enzyme has been named *dihydroxyacetone synthetase*. Again, for every three formaldehyde units assimilated, three five-carbon acceptor units are used and must therefore be regenerated. Thus, of the six three-carbon units produced as a result of the initial reaction, five are recycled to regenerate the acceptor, while the remaining one is used for the synthesis of cell material. The regeneration reactions involve one other enzyme characteristic of this pathway, *triokinase*, which is responsible for the phosphorylation of dihydroxyacetone.

The stoichiometry of the XMP pathway is:

$$3HCHO + 3ATP \rightarrow \text{glyceraldehyde-3-phosphate} + 2ADP + 2Pi$$

or, to compare directly with other pathways, when pyruvate is taken as the end product:

$$3HCHO + ATP + NAD(P)^+ \rightarrow \text{pyruvate} + ADP + Pi + NAD(P)H + H^+.$$

This compares very favourably with various versions of the RMP pathway (see Section 3.10), is considerably better than the L-serine pathway (see Section 3.11) and markedly more efficient than the RBP pathway (see Section 3.9) and serves to emphasize that the relatively slow growth of yeasts on methanol is most probably a reflection of their inefficient oxidative pathway (see Section 3.5). Should yeasts be seen in future as sources of single cell protein when grown on methanol, then genetic manipulation to introduce a more efficient oxidative route would seem to be essential.

## 3.13  Summary

In Sections 3.4 to 3.7 the ability of methylotrophs and carboxydotrophs to oxidize a wide variety of one-carbon substrates to generate ATP was examined. From this it is apparent that, despite the relatively limited number of substrates concerned, a variety of different mechanisms, particularly enzymic mechanisms and electron transport systems, are involved in different organisms growing on different substrates. Thus, although the usual oxidative route is via alcohol, aldehyde and acid to carbon dioxide with compounds feeding in at the appropriate point, a plethora of different variants are involved, the most unusual system being that found in yeasts.

One of the significant features of all the pathways is that they are linear, resulting in the complete oxidation of the substrate to carbon dioxide. This is in marked contrast to organisms using compounds containing two or more carbon atoms (with the exception of oxalic acid; see Section 4.2) which are oxidized via cyclic pathways, notably the TCA cycle (see Section 2.1) and, in the case of some two-carbon compounds, the dicarboxylic acid cycle (see Section 4.2). This serves to illustrate one of the basic metabolic differences between organisms growing on one-carbon substrates and substrates with three or more carbons, as discussed in Section 2.1. These differences are also apparent from the assimilatory routes. Organisms growing on compounds with three or more carbons produce three-carbon intermediates which may be used directly or anaplerotically carboxylated to oxaloacetate to provide the necessary precursors for growth (see Section 2.2). Organisms growing on one-carbon substrates must, however, use the substrate to generate these three-carbon compounds. This necessitates the presence of special pathways. As described in Sections 3.9 to 3.12 a wide variety of such pathways are used by various methylotrophs including the RBP pathway for autotrophic carbon dioxide fixation. These pathways, albeit three of them being very similar in outline, represent a splendid example of the adaptability of micro-organisms. Methylotrophs have clearly evolved along quite different routes to be able to utilize substrates, some of which are extremely abundant in the natural environment. Although they undoubtedly play an important role in natural ecosystems, particularly in the carbon cycle, it is probable that in the next decade our knowledge of their basic biochemistry will be applied in a wide range of biotechnological processes. These will include biomass production (not least because some obligate methylotrophs when growing on methane have also been shown to be able to fix gaseous nitrogen (Murrell and Dalton, 1983)), bioconversions in chemical manufacture, use in both detoxification processes and in the development of a range of biosensors.

In order to take full advantage of the potential of these organisms an important gap in our knowledge needs to be filled. To date little is known about the control of one-carbon metabolism including both fine control systems, to regulate flow of material through the oxidative and assimilatory pathways (likely to be mainly achieved through allosteric enzymes), and coarse control systems, which regulate enzyme levels. What little data is available does not allow any major insight or common patterns of control to be postulated at this stage (McNerny and O'Connor, 1980; Large, 1983). It is clearly important, particularly if these organisms are going to be 'improved' by genetic manipulation that control mechanisms are understood as soon as possible.

### Appendix to Chapter 3: Production of methane by methanogens

Anaerobes play a key role in the carbon cycle being responsible for much of the degradation of organic substances which lead to the eventual regeneration of carbon dioxide. As shown in Fig. 3.7 this process is now known to include three stages and three different groups of anaerobes. Group I organisms are known as *degraders* and effect the hydrolytic breakdown of polymers such as proteins, nucleic acids, and polysaccharides to produce, hydrogen, carbon dioxide, formate, acetate, ethanol, propionate, propanol, butyrate and butanol etc. Group II organisms are known as *acetogens* and further ferment ethanol, propionate, butyrate etc. to yield acetate, hydrogen and carbon dioxide as the major end products. Group III organisms are the *methanogens* and degrade acetate to yield methane and carbon dioxide.

The products of many of these reactions, particularly stage 1 reactions which are effected by a very wide variety of anaerobes such as *Clostridium*, *Bacteroides*, and

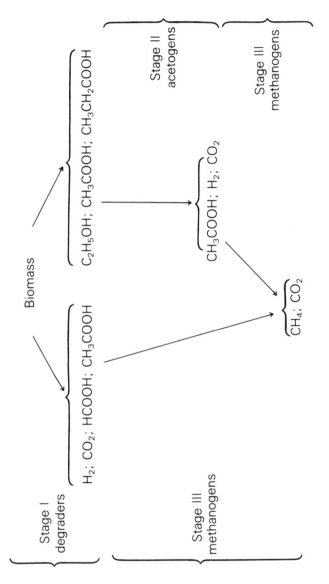

**Fig. 3.7** Stages in the anaerobic breakdown of organic substrates

*Ruminococcus*, are of great commercial and biotechnological importance. They include ethanol (as a fuel and beverage) and propanol and butanol (as solvents). Recently, increasing interest has been shown in the industrial applications of methanogens and it is appropriate to discuss these organisms in this appendix.

While methanogenic bacteria play a key role in the generation of methane in natural environments, the main reason for the current biotechnological interest in these organisms centres around the exploitation of their ability, as part of an assembly of bacterial anaerobes, to generate methane during the breakdown of domestic, agricultural and industrial wastes and effluents. Already some 26% of the fuel consumption in the United States is methane. Biotechnological generation of methane is seen as a major way of supplementing the natural geological sources of this gas. The process, known as *biogas production*, is already in widespread use but usually on a small scale (for example, as a heating source for farms when derived from pig, cattle or chicken manure) but the potential of such processes is enormous (Daniels, 1984). The potential will, however, only be realized when a more comprehensive knowledge of the physiology, biochemistry and molecular biology of methanogens is available. As will be seen, some aspects of these organisms, many of them peculiar to the group, are well documented but other aspects are only poorly understood. For a detailed treatment the excellent reviews by Large (1983), Vogel *et al.* (1984) and Zeikus (1983) should be consulted.

### A1   *The organisms*

Methanogens are now recognized as one of the groups of primitive bacteria (*Archaebacteria*) which include the extreme halophiles and thermoacidophiles. These organisms lack a peptidoglycan component in their cell wall and have certain other characteristics, particularly with respect to their mechanism of protein synthesis, which distinguish them from the majority of bacteria (the *Eubacteria*). They form a very restricted and specialized group of bacteria which live in waterlogged soils, the gut of animals, sewage sludge, rotting vegetation and aquatic sediments. The taxonomy of methanogens is gradually being rationalized and to date three orders of strict (obligate) methanogens have been recognized. These are the Methano-bacteriales, including species of *Methanobacterium* and *Methanobrevibacter*, the Methanoc-occales including species of *Methanococcus* and the Methanomicrobiales including species of *Methanomicrobium*, *Methanogenium*, *Methanospirillum* and *Methanosarcina*.

Methanogens are mostly non-motile and represent some of the largest bacterial cells so far isolated. They are amongst the strictest anaerobes, $0.01 \text{ mg l}^{-1}$ of oxygen being totally inhibitory. Like most anaerobes they lack superoxide dismutase and catalase and hence cannot remove superoxide radicals or hydrogen peroxide which result from oxygenic metabolism. Nitrate and sulphate are also inhibitory, while ammonia and carbon dioxide are essential for growth, and hydrogen sulphide generally has a stimulatory effect. Cysteine can replace hydrogen sulphide, but amino acids and peptides cannot substitute for ammonia. Until recently all methanogens were believed to grow only very slowly, the doubling time being in excess of 10 h even in optimal conditions. However, some recent isolates (such as *Methanococcus jannaschii*) have much more rapid growth rates with doubling times of little more than 30 min.

The majority of methanogens grow best using carbon dioxide and hydrogen as their growth substrates, the energy necessary for growth being derived from the reduction of the carbon dioxide to methane by hydrogen. Some methanogens utilize formate instead of hydrogen and carbon dioxide. A small number will utilize either formate or methanol or methylamine. The most versatile species, in terms of the range of substrates utilized, are strains of *Methanosarcina* since they will use hydrogen, carbon dioxide, methanol, methylamine and acetate, the latter being a substrate most other species isolated to date will not utilize.

Any organism which grows using carbon dioxide as its apparent growth substrate is, strictly speaking, an *autotroph*. However, methanogens are not placed in this category since their utilization of carbon dioxide is quite different from other autotrophs. There is no evidence for the assimilation of carbon dioxide via the RBP pathway. Indeed the crucial enzyme for autotrophic carbon dioxide fixation via the RBP pathway, ribulose bisphosphate carboxylase, is totally absent.

## A2  *The methanogenic process*

The mechanism of methane generation and of concomitant ATP production has been examined in detail in the past few years and remarkable progress has been made. It is already clear that methanogenesis involves several unusual reactants which are worth listing straight away. These are:

(i) *tetrahydromethanopterin*, which is a carrier of one-carbon units and therefore, functionally, resembles tetrahydrofolic acid, but which is apparently unique to methanogens;

(ii) *co-enzyme $F_{420}$*, which is a deazoflavin, very different in structure to FMN or FAD with a long side chain comprising a lactyl group and two glutamyl residues which acts as an electron carrier during methanogenesis;

(iii) a *carbon dioxide reduction factor*, which has not been fully characterized;

(iv) *co-enzyme M*, which is 2-mercaptoethane sulphonic acid ($HSCH_2CH_2SO_3^-$) and is important in the terminal steps of methanogenesis, again being unique to methanogens;

(v) a nickel containing yellow *factor $F_{430}$*; and

(vi) *factor III*, neither of whose roles in methanogenesis are not yet established.

From the above list it will already be clear that the biochemistry of methane formation involves unique components and reactions, characteristic of this group of organisms. The fact that the only usual component of electron transport systems which has been found in methanogens is cytochrome *b* simply serves to emphasize this point.

The whole process of methane formation appears to occur with all intermediates remaining tightly bound to the carriers. The first step involves a carrier which is probably tetrahydromethanopterin (Vogels *et al.*, 1981) and carbon dioxide is converted to carrier-bound formate. Subsequently, as illustrated in Fig. 3.8 the carrier-bound formate is successively reduced to formaldehyde, methanol and methyl intermediates before the terminal step in which methane is released. It is likely but not definite that coenzyme F $(CoF)_{420}$ in its reduced form, directly reduced by a *hydrogenase* system, acts as the main reducing agent during this process. The best understood step is the terminal one in which coenzyme M is involved. It appears likely that the one-carbon unit is transferred to CoM as the methanol derivative. This is then reduced to a methyl derivative and in the final step methane is released. The terminal step is catalysed by *methyl CoM reductase* and involves three components, one of which is yellow and may be $F_{430}$. This reaction may be summarized as:

$$CH_3-S-CoM + H_2 + ATP \rightarrow CH_4 + CoM - SH + ADP + Pi.$$

The purpose of the overall process of methanogenesis is to generate ATP yet the terminal reaction requires ATP. Where then is ATP generated in this process? To

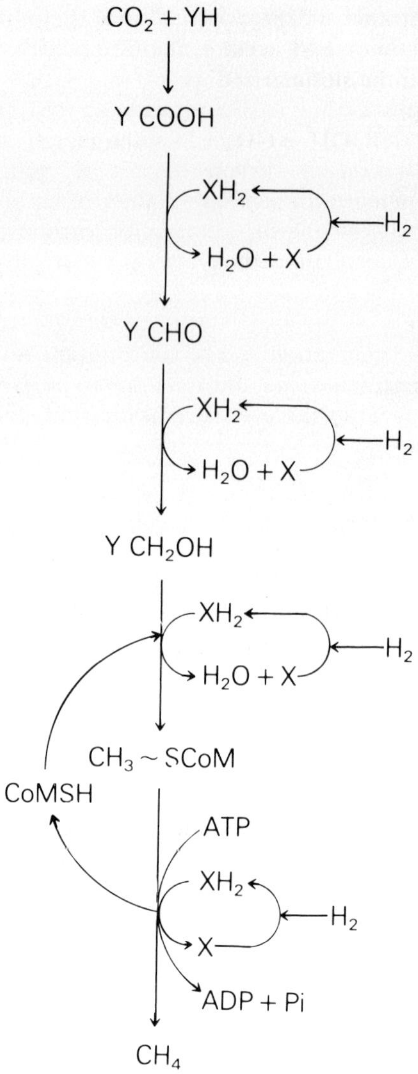

**Fig. 3.8** The pathway of methane production. Y is a carrier, probably tetrahydro-methanopterin. $XH_2$ represents one or more electron carriers which can be directly or indirectly reduced by hydrogen and may include $CoF_{420}$ and/or $F_{430}$

date the process of ATP production is only poorly understood. Speculation would be unfruitful and may only serve to confuse. It is clear that this is a major gap in our knowledge of these organisms which must be filled.

The overall process of methanogenesis using carbon dioxide and hydrogen as the substrates can be summarized as:

$$CO_2 + 4H_2 \rightarrow CH_4 + 2H_2O.$$

When formate is used as the growth substrate, the pathway of methane generation is not dissimilar to that illustrated in Fig. 3.8 in that it is probably used as the direct reducer of $F_{420}$ with the release of carbon dioxide which will then be used as illustrated. This process can be summarized as:

$$4HCOOH \rightarrow CH_4 + 3CO_2 + 2H_2O.$$

Rather less is known about the utilization of methanol and acetate but the processes involved can be summarized as:

$$4CH_3OH \rightarrow 3CH_4 + CO_2 + 2H_2O$$

and

$$CH_3COOH \rightarrow CH_4 + CO_2.$$

These simple equations mask a complex process which, in the case of methanol, involves initial oxidation of some methanol in order to generate the reducing power to subsequently reduce methanol to methane. Thus the reactions are better written as:

$$CH_3OH + H_2O \rightarrow CO_2 + 3XH_2$$
$$3CH_3OH + 3XH_2 \rightarrow 3CH_4 + 3H_2O$$

which when summated give the previous summary equation. It is likely that in the reduction of methanol to methane the methyl group of methanol is transferred directly to CoM.

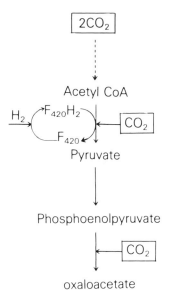

**Fig. 3.9** Outline of pathway for generation of three and four carbon precursors for biosynthesis in methanogens

### A3  *The synthesis of cell materials in methanogens*

As has already been established, methanogens do not fix carbon dioxide autotrophically. How then do these organisms convert carbon dioxide to cellular materials? While some progress has been made, the initial reaction in which hydrogen and carbon dioxide are used to generate acetate remains unresolved. It is possible, indeed likely, that similar one-carbon carriers to those involved in methanogenesis are used and that two of these, or possibly one of these and a one-carbon derivative of vitamin $B_{12}$, which is present in methanogens in enormous quantities, react together to form acetyl CoA. This, as shown in Fig. 3.9, is carboxylated to pyruvate which in turn is carboxylated to oxaloacetate. Thus the three-carbon and four-carbon compounds required for biosynthesis can be generated. These unique reactions, which do not occur in aerobes, include a *pyruvate synthetase*, similar to that in photosynthetic bacteria. This enzyme, which synthesizes pyruvate by the carboxylation of acetyl CoA, has been demonstrated in *Methylosarcina barkeri* and depends on $F_{420}$ for the required reducing power (in photosynthetic bacteria reduced ferredoxin is required). If these reactions are confirmed then the initial formation of acetyl CoA will need to be explained in order to be able to fully account for how these organisms provide themselves with their biosynthetic requirements.

### A4  *Summary*

It is clear from the above discussion that the biochemistry of methanogens is fascinating but, as yet, incompletely understood. In particular the mechanism of ATP generation and the method of acetate production from carbon dioxide are the two main outstanding problems. With the great interest being shown in the biotechnological exploitation of these organisms it is likely that knowledge of methanogens will rapidly expand.

# Chapter 4

## Aerobic microbial growth on two-carbon substrates

### 4.1 Scope and rationale

There is a much greater variety of two-carbon compounds than one-carbon compounds, consequently a greater variety of organisms are able to utilize such substrates and they do so using a greater variety of both dissimilatory, oxidative energy-yielding pathways, and assimilatory, biosynthetic pathways. However, an examination of the substrates themselves and the pathways by which they are metabolized permits certain broad generalizations to be made. These generalizations are based on the oxidation/reduction states of the substrate which include three categories: substrates with a H:O ratio of 2:1 or more (for example, ethanol ($C_2H_5OH$) and acetate ($CH_3COOH$)); substrates with a H:O ratio of less than 2:1 but more than 0.6:1 (for example, glycollate ($CH_2OHCOOH$) and glyoxylate ($CHOCOOH$)); and very highly oxidized substrates with H:O ratios below 0.6:1 (for example, oxalate ($COOH)_2$). In general, growth on two-carbon substrates has aroused less biotechnological interest than growth on one-carbon substrates. This is not to say this lower level of interest will continue, since many two-carbon compounds are generated in natural ecosystems and as industrial byproducts, and they may be potentially useful substrates for biomass production.

### 4.2 Oxidative pathways for ATP generation

Most two-carbon substrates, which have been used in studies on oxidative, energy-yielding, metabolism have H:O ratios of 2:1 or more. By and large they are oxidized to form acetyl CoA which is further oxidized via the TCA cycle. For example:

$$C_2H_5OH \xrightarrow[X \quad XH_2]{} CH_3CHO \xrightarrow[X \quad XH_2]{CoASH} CH_3CO{\sim}SCoA \longrightarrow TCA\ cycle$$

ethanol        acetaldehyde        acetyl CoA

Thus, with respect to oxidative metabolism, such substrates are essentially utilized as in all other organisms (see Section 2.1). As will be shown this is not the case, when anaplerotic routes are considered (see Section 4.3).

More highly oxidized substrates with H:O ratios less than 2:1, but more than 0.6:1 are oxidized via two known routes. Firstly they may be reduced to the level of acetate which is then oxidized via the TCA cycle (that is, a reduction/oxidation route). This unusual route occurs in *Paracoccus denitrificans* and is a very interesting pathway (see Fig. 4.1) that was discovered because the ability of this organism to oxidize glycollate or glyoxylate was blocked by the classical TCA cycle inhibitors, monofluoroacetate, malonate, and arsenite.

In some respects the pathway shown in Fig. 4.1 is relatively unique. The key reaction is the condensation of glycine (2C) and glyoxylate (2C) to form hydroxy-aspartate (4C) catalysed by *hydroxypyruvate synthetase*. The amino acid is then deaminated, to its keto-acid equivalent, decarboxylated to phosphoenolpyruvate and further metabolized via pyruvate to a second decarboxylation step which yields acetate. This is then oxidized via the TCA cycle. Clearly, the pathway generates reduced electron acceptors (in its own right as well as via the oxidative steps of the TCA cycle). These reduced acceptors are reoxidized via the normal processes of electron transport and ATP is generated. The unusual feature of the pathway is that, as part of the route in which acetate is generated, a C2–C2 condensation reaction occurs which produces a four-carbon compound. Hence the pathway fills not only an oxidative, but also an anaplerotic role.

Overall, using glyoxylate as the example and assuming normal mechanisms of electron transport, this pathway yields for the complete oxidation of two molecules of glyoxylate some 15 ATP (that is, an ATP:C ratio of 3.75) which compares to the 12 ATP generated from the complete oxidation of one molecule of acetate (that is, an ATP:C ratio of 6). The lower energy yield is a reflection of the more highly oxidized status of the substrate.

The second and more usual pathway for the oxidation of substrates such as glycollate and glyoxylate (and glycine derived from glyoxylate by deamination) is known as the *dicarboxylic acid (DCA) cycle* (see Fig. 4.2). This has been shown to occur in coliforms, pseudomonads and many other groups of bacteria utilizing these substrates. Its initial discovery resulted from the observation that, whilst monofluoroacetate markedly inhibited oxidation of glyoxylate in coliforms, the other classical TCA cycle inhibitors did not. This paradox was resolved, following observations that malate was the initial product of glyoxylate oxidation and its formation depended on the induction, prior to growth on glyoxylate, of the enzyme *malate synthetase*. This enzyme was subsequently shown to be inhibited by monofluoroacetate.

After its formation, malate is oxidized to oxaloacetate and this is followed by two decarboxylation steps to yield, firstly, pyruvate and subsequently acetyl CoA. As shown in Fig. 4.2 acetyl CoA acts as the initial acceptor of glyoxylate and is regenerated as a result of the last reaction of the cycle. Overall, therefore, the DCA cycle accounts for the complete oxidation of one molecule of glyoxylate to carbon dioxide and the generation of two reduced electron acceptors and one ATP. Assuming normal electron transport processes, the oxidation of glyoxylate will therefore yield 7 ATP (that is, an ATP:C ratio of 3.5).

It is interesting to compare the TCA and DCA cycles. The former accounts for the complete oxidation of acetyl CoA with the obligate participation and regeneration of a keto-acid, whilst the latter accounts for the complete oxidation of a keto-acid (glyoxylate) with the obligate participation and regeneration of acetyl CoA. However,

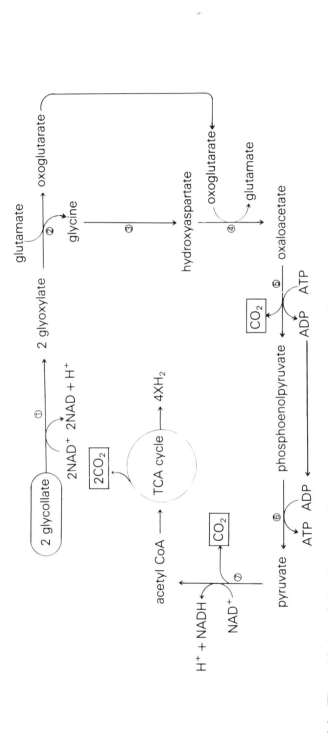

**Fig. 4.1** The oxidation of glycollate and glyoxylate in *Paracoccus denitrificans*; ① glycollate dehydrogenase; ② glyoxylate:glutamate transaminase; ③ hydroxyasparatate synthetase; ④ oxaloacetate:glutamate transaminase; ⑤ PEP-carboxykinase; ⑥ pyruvate kinase; ⑦ pyruvate dehydrogenase

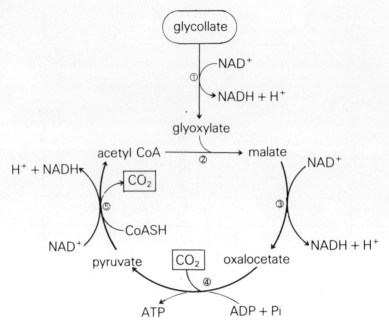

**Fig. 4.2**   The dicarboxylic acid cycle; ① glycollate dehydrogenase; ② malate synthetase; ③ malate dehydrogenase; ④ pyruvate carboxylase; ⑤ pyruvate dehydrogenase

the role of the two pathways is identical, namely, to oxidize a two-carbon substrate as part of the ATP generating process and to provide the biosynthetic precursors for growth (acetyl CoA, pyruvate, malate and oxaloacetate). Thus, both pathways are *amphibolic* and to satisfy the demand for biosynthetic precursors an *anaplerotic* pathway is required (see Section 4.3). It should also be noted that organisms utilizing the DCA cycle to oxidize substrates such as glyoxylate will still have a requirement for at least a partial TCA cycle. This is because they need to produce the five-carbon oxoglutarate as the precursor of glutamic acid which plays a key and central role in amino acid metabolism (see Fig. 2.1 and Section 2.2).

Relatively few studies have been carried out on the oxidative pathways used to generate ATP from highly oxidized two-carbon compounds such as oxalate. This compound is used in considerable quantities as a rust remover prior to paint spraying in the car industry, and presents something of a disposal problem to the industry because it is fairly toxic. Consequently, interest has been shown in using micro-organisms for its detoxification. Needless to say such a highly oxidized substrate is not a good growth substrate, the potential for ATP generation being small. However, some pseudomonads including *Pseudomonas oxalaticus* can grow an oxalate, albeit relatively slowly and inefficiently.

The pathways involved in oxalate utilization represent a break point between the normal cyclical pathways for oxidating of two-carbon substrates (TCA or DCA cycles) and the linear oxidative route involved in one-carbon metabolism (see Section 3.4). The oxidative route has more in common with oxidation of one-carbon than two-carbon substrates since it is linear. On the other hand, as discussed in Section 4.3, the biosynthetic pathways are more typical of growth on two-carbon substrates.

The linear oxidative route is simple (see Fig. 4.3). The important enzyme is

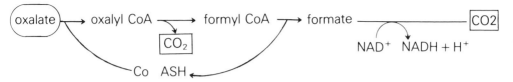

**Fig. 4.3** Linear oxidative pathway for oxidation of oxalic acid

*formate dehydrogenase*, which is also involved in oxidation of one-carbon substrates by methylotrophs. When the reduced NADH produced by this reaction is oxidized, 3 ATP result thus giving an ATP:C ratio of 1.5, a very low yield reflecting the nature of the highly oxidized substrate.

It must be noted that, unlike other pathways for oxidation of two-carbon compounds this pathway does not involve three-carbon and four-carbon biosynthetic precursors. Thus, strictly speaking, no anaplerotic, pathway is required but, as in methylotrophs, a mechanism of generating three-carbon and four-carbon precursors is needed.

## 4.3 Biosynthetic and anaplerotic pathways in organisms growing on two-carbon substrates

As indicated earlier (see Section 4.2) the majority of two-carbon compounds are oxidized via the TCA cycle. This necessitates an anaplerotic pathway to top up levels of TCA cycle intermediates required for biosynthetic precursors. Since there are few organisms which have evolved mechanisms of carboxylating two-carbon compounds to pyruvate or phosphoenolpyruvate which can be further carboxylated to oxaloacetate or malate, some other anaplerotic mechanism of converting the two-carbon substrates to four-carbon precursors is required.

As already discussed, the unusual reduction/oxidation route of *Paracoccus denitrificans* fulfils this role (see Section 4.2) but the more usual oxidative routes found in other organisms (that is, the TCA and DCA cycles) require a separate anaplerotic pathway. The nature of the pathway used, again reflects the H:O ratios of the substrate and the oxidative route. For compounds with a H:O ratio of 2:1 or more the *glyoxylate cycle* is used to top up levels of TCA cycle intermediates, while for compounds with lower H:O ratios which are oxidized via the DCA cycle the *glycerate pathway* is used. This includes the very highly oxidized oxalic acid which, as recorded earlier, is used biosynthetically like other two-carbon compounds even though it is oxidized as though it were a one-carbon compound.

The *glyoxylate cycle* (see Fig. 4.4) involves the normal route of entry into the TCA cycle. Thus acetyl CoA condenses with oxaloacetate to form citrate, which is further metabolized via TCA cycle reactions to isocitrate. At this point the two successive oxidative decarboxylations of the TCA cycle, which produce first oxoglutarate and then succinate, are bypassed. Isocitrate (6C) is split by *isocitrate lyase* to produce succinate (4C) and glyoxylate (2C). Glyoxylate then condenses with a second molecule of acetyl CoA (2C) to form malate (4C) in a reaction catalysed by *malate synthetase*. The malate is then oxidized to regenerate the original oxaloacetate. Overall the pathway,

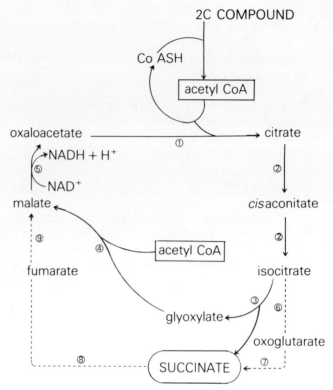

**Fig. 4.4** The glyoxylate cycle. Reactions common to the TCA cycle—① citrate condensing enzyme; ② aconitase; ⑤ malate dehydrogenase. Reactions specific to the glyoxylate cycle—③ isocitrate lyase; ④ malate synthetase. Reactions specific to the TCA cycle—⑥ isocitrate dehydrogenase; ⑦ oxoglutarate dehydrogenase; ⑧ succinic dehydrogenase; ⑨ fumarate dehydrogenase

therefore, results in the net production of one four-carbon compound (succinate) thereby fulfiling the necessary anaplerotic function:

$$2CH_2COOH + NAD^+ \longrightarrow \begin{matrix} CH_2COOH \\ | \\ CH_2COOH \end{matrix} + NADH + H^+.$$

acetate                             succinate

The *glyceric acid pathway* (see Fig. 4.5) results in the production of oxaloacetate to top up the TCA and DCA cycles, but this is achieved in a rather different way to the glyoxylate cycle. Instead of the C2–C2 condensation of the glyoxylate cycle, two glyoxylate molecules are involved in a reaction which yields carbon dioxide and the three-carbon tartronic acid semialdehyde (TAS). This reaction is catalysed by an enzyme unique to the pathway, *glyoxylate carboligase*. Further metabolism of TAS is via its reduction to glyceric acid by a second enzyme characteristic of this pathway, *TAS reductase*. The glyceric acid is phosphorylated and metabolized via normal glycolytic reactions to phosphenolpyruvate or pyruvate, either of which is then carboxylated to form oxaloacetate—the route depending on which carboxylating enzyme, *pyruvate carboxylase* or *PEP carboxykinase* is present in the organism. Overall the sequence of

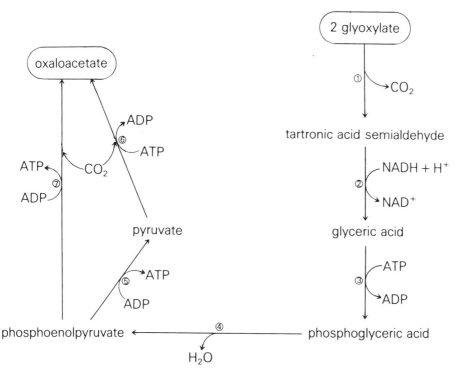

**Fig. 4.5** The glycerate pathway; ① glyoxylate carboligase; ② tartronic acid semialdehyde reductase; ③ glycerate kinase; ④ enolase; ⑤ pyruvate kinase; ⑥ pyruvate carboxylate; ⑦ phosphoenolpyruvate carboxykinase. NB: *Either* reactions ⑤ and ⑥ *or* ⑦ would lead to the formation of oxaloacetate in a given organism

events can be summarized as:

$$C2 + C2 \longrightarrow C3 + CO_2 \longrightarrow C4$$

or as:

$$2\ \underset{\text{glyoxylate}}{CHOCOOH} + NADH + H^+ \underset{\hspace{1.5em}}{\overset{CH_2COOH}{\underset{COCOOH}{|}}} + NAD^+ + H_2O$$
oxaloacetate

As mentioned above, oxalic acid is also assimilated via this pathway. Since *Pseudomonas oxalaticus* assimilates formate via the autotrophic route of the RBP pathway after oxidizing it all to formate, it might be expected that oxalate would be oxidized to formate and then fixed as carbon dioxide. However, the labelling patterns using cells grown on [14]C-formate and [14]C-oxalate were found to be very different, with tartronic acid semialdehyde and glyceric acid as early labelled intermediates from [14]C-oxalate indicating glyceric acid pathway activity. It was subsequently demonstrated that oxalyl CoA is reduced to form glyoxylate:

$$\text{oxalyl CoA} + NADH + H^+ \rightarrow \text{glyoxylate} + NAD^+ + CoASH$$

which feeds into the glycerate pathway.

## 4.4   Summary

Microbial growth on two-carbon substrates involves a variety of oxidative and anaplerotic pathways according to the initial oxidation/reduction status of the substrate. Most two-carbon compounds are metabolized to yield acetate which is then oxidized via the TCA cycle. Since this pathway is amphibolic, providing biosynthetic precursors as well as serving as the terminal oxidative pathway, an anaplerotic pathway is required. This is generally the glyoxylate cycle. For more highly oxidized substrates the DCA cycle is the oxidative route and the anaplerotic pathway is the glycerate pathway. The highly oxidized oxalic acid is oxidized via a linear pathway, rather like that for the oxidation of one-carbon substrates, but assimilated via a two-carbon route, the glycerate pathway.

More work has been done on the control mechanisms involved in growth using two-carbon compounds than for one-carbon compounds. In general these studies indicate that the enzymes required to convert a given substrate to acetyl CoA or glyoxylate to enter into the TCA or DCA cycles are induced by the presence of the substrate as *sole* carbon substrate. This is certainly true of specific enzymes of the anaplerotic glyoxylate cycle and glycerate pathways.

Particular interest has centred on the role of the enzymes isocitrate lyase and malate synthetase in one-carbon and two-carbon metabolism. Isocitrate lyase is a key enzyme in the L-serine pathway found in type II facultative methylotrophs which, besides being able to grow on one-carbon substrates can also utilize a wide range of two-carbon compounds. When growing on most two-carbon substrates the enzyme is essential in the glyoxylate cycle. It is interesting that the role of the enzyme, even though it catalyses an identical reaction in each case, is rather different. In the L-serine pathway, it is part of a biosynthetic pathway and is essential to the mechanism of regenerating an acceptor unit (glycine) for a one-carbon derivative of THF. On the other hand in the glyoxylate cycle it has a crucial, anaplerotic role.

The enzyme malate synthetase enjoys two even more markedly contrasting roles in growth of two-carbon compounds. In growth on glyoxylate, it is active in the DCA cycle oxidative pathway while in growth on acetate, or compounds converted to acetate, it has an anaplerotic function in the glyoxylate cycle. This remarkable dual use of such enzymes raises the question of how their induction is controlled. This question has not been satisfactorily answered as yet, but it is clear that, at least in the case of malate synthetase, the two enzymes involved are quite distinct gene products, one of which is induced by substrates oxidized via the DCA cycle, the other being induced by substrates oxidized via the TCA cycle and assimilated by the glyoxylate cycle.

Similarly the control of metabolic pathways in *Pseudomonas oxalaticus* is also of considerable interest. In particular, the question of why and how ribulose-1,5-bisphosphate carboxylase is induced by growth on formate, which is oxidized via formate dehydroxygenase to carbon dioxide and fixed via the carboxylase, while oxalate, which is also oxidized via formate dehydrogenase to carbon dioxide causes ribulose-1,5-bisphosphate carboxylase repression and induces glycollic dehydrogenase, glyoxylate carboligase and TAS reductase, is an interesting one. What is clear is that the dual uses of enzymes in one-carbon and two-carbon metabolism and, in the case of oxalate, the fascinating choices of metabolic route, both further serve to indicate the versatility of micro-organisms in growing on various substrates.

# Chapter 5

## *Aerobic microbial growth on selected substrates with more than two carbons*

### 5.1  Scope and rationale

As indicated in Chapter 2, Sections 2.1 to 2.3 all compounds with three or more carbons are metabolized via routes which are found in nearly all living organisms and it is not intended that these routes should be considered here. Essentially, such substrates are converted, usually via fairly direct routes, to intermediates of glycolysis prior to oxidation via the TCA cycle. Anaplerosis is via carboxylation of three-carbon intermediates. However, two types of compounds which are used as microbial growth substrates are of particular biotechnological interest. These are hydrocarbons, other than methane, and aromatic compounds. Some specialist aspects of their utilization as growth substrates are considered in this chapter.

### 5.2  Oxidation and assimilation of aliphatic hydrocarbons

The ability of micro-organisms to degrade hydrocarbons has long been recognized and was first reviewed by Zobell (1950). Initial studies were hampered by the lack of availability of pure substrates. This resulted in peculiar growth and respiratory patterns which were shown to be due to the organisms oxidizing, and presumably utilizing as growth substrates, one impurity followed by a decrease in growth and oxygen uptake while a new set of enzymes were induced to utilize a second impurity, and so on until all the impurities had been utilized. Only then was the hydrocarbon used as substrate by which time, in batch culture, other nutrients were limiting.

   Interest in exploiting organisms that can degrade hydrocarbons is now quite intense from several points of view. First, such organisms may be used for biomass production, using waste oils or other hydrocarbon waste products as substrates. Second, they are being actively examined for their possible use in overcoming problems of oil pollution, particularly for the dispersion of oil spills. Third, they are being examined for their potential to release oils from oil-containing substrate specially since, using conventional techniques, it is generally impossible to recover more than 70% of the oil from an oil well. In view of these biotechnological applications it is important to examine the range of organisms capable of degrading hydrocarbons and how they do so. Moreover, hydrocarbon-degrading micro-organisms, besides showing potentially beneficial characteristics, also cause problems because that they can grow in oil/water interfaces and in oil/water emulsions thereby causing biodeterioration problems in mechanical systems due to oil degradation.

Many bacteria, yeasts and filamentous fungi have been shown to be able to oxidize and assimilate aliphatic hydrocarbons. They include species of *Corynebacterium*, *Pseudomonas*, and *Candida*. In general the short-chain hydrocarbons, while being oxidized, are not used as sole growth substrates, but hydrocarbons with a chain length of eight (for example *n*-octane) or more support good growth. As a broad generalization, saturated hydrocarbons are utilized more readily than unsaturated ones while straight-chain hydrocarbons are preferred to branched ones and even-numbered hydrocarbons to odd-numbered ones.

The oxidative route (see Fig. 5.1) basically follows that for growth on methane:

$$hydrocarbon \rightarrow alcohol \rightarrow aldehyde \rightarrow fatty\ acid$$

with the resulting fatty acid being further metabolized usually, via *β-oxidation*, to acetyl CoA which is terminally oxidized via the TCA cycle. The *mono-oxygenase* system responsible for hydrocarbon oxidation to the corresponding alcohol and the *alcohol* and *aldehyde dehydrogenases* are similar to those used in methane conversion to methanol, formaldehyde and formate. Two different mechanisms of associated electron transport for the mono-oxygenase have been described, one dependent on cytochrome $P_{450}$ in *Corynebacterium* sp. and the other in *Candida* sp. which is dependent on rubredoxin. The dehydrogenases have been shown to be NAD-linked (Wyatt, 1984).

In some cases *diterminal oxidation* can occur (see Fig. 5.1). This yields dicarboxylic fatty acids which must be subsequently oxidized to acetyl CoA via *α-ω oxidation*, oxalic acid being an additional end product. The subsequent fate of the oxalic acid is not yet clear. Some species may use *subterminal* oxidation in which a carbon other than one of the terminal carbons is the site of mono-oxygenase attack. Once again the end product is acetyl CoA which is further oxidized via the TCA cycle. Rather less is known about the oxidation of alkenes (unsaturated hydrocarbons) or branched chain hydrocarbons, but it is clear that mono-oxygenases are again involved. As indicated, the end product of the oxidative pathway for hydrocarbon oxidation is acetyl CoA which is subsequently further oxidized via the TCA cycle. Thus much ATP will be generated when hydrocarbons are oxidized, not least because *β* or *α-ω* oxidation also generate reduced nucleotides.

When it comes to the route for the production of three-carbon and four-carbon precursors for biosynthesis, in effect the organisms concerned can be considered to be growing on acetate as this is the end product of the initial stage of oxidation. Not suprisingly, therefore, organisms growing on aliphatic hydrocarbons contain *isocitrate lyase* and *malate synthetase*, indicating that the *glyoxylate cycle* is used as the anaplerotic pathway.

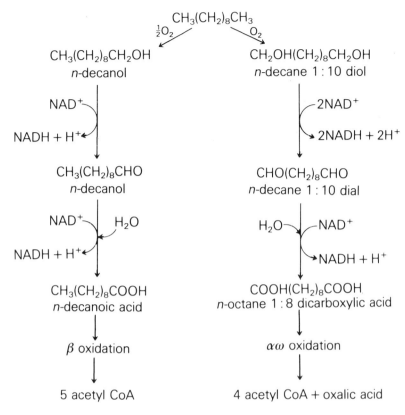

**Fig. 5.1** Oxidative routes for the degradation of long-chain hydrocarbons illustrated by *n*-decane; monominal oxidation is shown on the left and ditermal oxidation on the right of the figure

A fascinating aspect of the growth of micro-organisms on aliphatic hydrocarbons is the actual uptake of these insoluble substances. The indications are that the organisms secrete specific surfactant agents which emulsify the hydrocarbons into a form which aids transport across the cell membrane. It is also known that, when growing on hydrocarbons, micro-organisms have a much higher lipid content in their membranes. It may be that the combination of hydrocarbon emulsions coming into contact with highly lipidic membranes is sufficient to account for hydrocarbon uptake, but it is equally possible that, as more work is done, special uptake pathways will be unravelled.

## 5.3   Oxidation and assimilation of aromatic substances

The ability of micro-organisms to degrade aromatic substances has been examined in a wide variety of organisms using an enormous range of compounds. Biotechnological interest in such degradations is largely from the point of view of utilizing immobilized enzymes or cells in specific bioconversions, particularly to catalyse reactions which otherwise may be difficult or inefficiently carried out by chemical means (for instance, in the synthesis of drugs or other desirable end products). A second area of potential application is in the biodegradation and hence clean-up of potentially toxic effluents containing aromatic substances. Similarly, persistent pesticides might be broken down and hence detoxified by using methods based on a knowledge of the degradation of aromatic compounds by micro-organisms.

The variety of aromatic compounds utilized by one micro-organism or another is vast and only a very brief treatment of important principles is attempted. For a comprehensive account the reader is referred to the detailed review by Dagley (1978). The metabolism of various aromatic compounds takes place via a wide range of mechanisms initially involving oxygenation and hydroxylation reactions. As by now may be expected, the oxygenation often involves mono-oxygenase activity. It is impossible here to give details since variation is very great. However, the crucial reaction for degradation of these substances is the opening of the benzene ring and there are only two main routes for this. Most aromatic substances which can be completely oxidized are initially converted to catechol or related compounds subjected to either *meta* or *ortho* fission. Further metabolism of the products of these reactions leads to the generation of intermediates of the TCA cycle, usually succinyl CoA and acetyl CoA. The latter is oxidized via the TCA cycle to generate ATP while the former can act as an anaplerotic top-up for the cycle. Thus the pathways involved in utilization of aromatic substrates yield end products which enable both requirements for growth to be satisfied.

## 5.4  Summary

Only a very brief and general treatment of aerobic microbial growth on compounds containing three or more carbon atoms has been attempted. This is because of the variety of such compounds and the fact that after one or two initial reactions the vast majority of them are oxidized via glycolysis and the TCA cycle (see Fig. 2.1). This route involves, as intermediates, three-carbon compounds which can be carboxylated to give four-carbon intermediates of the TCA cycle and thereby satisfy the provision of the biosynthetic precursors necessary for growth. Thus, what has been described earlier, as the 'normal' pathways of metabolism which represent the unifying theme of biochemistry, can operate for the majority of substrates with three or more carbons. Even the more specialized hydrocarbons and aromatic substances are metabolized according to the same principles (that is, conversion to intermediates of a small number of pathways which feed into central metabolic pathways.)

# SECTION III

# Culturing Micro-organisms for Production

## P. F. Stanbury

## Assumed concepts

Basic microbial biochemistry.
The structure and taxonomy of micro-organisms.
Basic algebra and calculus.

*Useful books*
Demain, A. L. and Solomon, N. A. (1986) Manual of Industrial Microbiology and Biotechnology. Washington, American Society for Microbiology.
Riviere, J. (1977) Industrial Applications of Microbiology (Translated and edited by Moss, M. O. and Smith, J. E.). Surrey University Press.
Stanbury, P. F. and Whitaker, A. (1984) Principles of Fermentation Technology, Oxford, Pergamon Press.

# Chapter 6

# *Products from micro-organisms*

## 6.1 Introduction

The biochemical diversity of micro-organisms, illustrated by their ability to grow on a variety of substrates and to produce a wide range of products, has resulted in their commercial exploitation by the fermentation industry. Microbiologists tend to interpret 'fermentation' as any process for the production of a product by the mass culture of micro-organisms, whereas biochemists use the term in a far more strict sense. The true biochemical meaning of the word is 'an energy-generating process in which organic compounds act as both electron donors and terminal electron acceptors' (see Section 1.1). However, in this section 'fermentation' will be used in its broader, microbiological sense. In recent years the advances in molecular biology and genetic manipulation have given the well-established fermentation industry the opportunity to initiate new processes and to improve existing ones. These developments have been hailed as the birth of a new era but it should be borne in mind that the exploitation of these major advances depends upon the techniques of large-scale culture that have been built up by the industry over many years. The objectives of this section are: first, to introduce the reader to the range of commercial fermentation products, and second, to concentrate on the biological principles underlying these processes.

## 6.2 The production of microbial biomass

The most obvious microbial product of commercial significance is probably microbial biomass (the microbial cells themselves). Yeast cells, used in the baking industry, have been produced commercially since the early 1900s, and yeast was also produced as human food in Germany during the First World War. These early biomass processes were aerobic and, as discussed below, incorporated a considerable degree of process control. Although the production of bakers' yeast by fermentation has continued, the First World War production of microbial biomass as human food or animal feed was a relatively isolated event and such systems were not investigated in any great depth until the 1960s. At that time the biochemical diversity of micro-organisms was exploited by the utilization of a wide variety of organisms capable of growing on a range of carbon sources, including hydrocarbons for the production of microbial biomass termed 'single cell protein' or SCP. Very few of the processes investigated at that time came to fruition due to economic and political difficulties. British Petroleum was probably the early leader in this field and developed its technology to the extent of constructing a commercial plant in Sardinia for the production of yeast biomass from *n*-alkanes. However, BP was unable to convince the Italian government of the toxicological safety of its product and the factory never went into production. Most of the other processes were casualties of the Middle East wars and the subsequent escalation in petroleum product prices. However, Imperial Chemical Industries perservered with their plans for the production of bacterial biomass (*Methylophilus methylotrophus*) from methanol and eventually established the Pruteen process for the production of high grade protein as an animal feed. The process utilizes continuous culture (see Chapter 7) on an enormous scale (1500 m³) and is an excellent example of the application of good engineering to the design of a microbiological process. The economics of the production of SCP as animal feed are still marginal, but the technology developed by ICI during the Pruteen project has established the company as a world leader in biotechnology. ICI are capitalizing on their expertise in large-scale continuous culture by collaborating with Rank Hovis MacDougall on a process for the production of fungal biomass from carbohydrates to be marketed as human food, and the economics of this process should prove more attractive.

## 6.3  The production of microbial products

The exploitation of the biosynthetic capabilities of micro-organisms has resulted in the establishment of a very large number of commercial processes. The growth of a micro-organism may result in the production of a range of metabolites, but the predominant type of metabolite synthesized depends upon the nature of the organism, the cultural conditions employed and the growth rate of the producing culture. If a micro-organism is introduced into a nutrient medium which supports its growth, the inoculated culture will pass through a number of stages and the system is termed batch culture (see Chapter 7). Initially, growth does not occur, and this period is referred to as the *lag phase* and may be considered a period of adaptation. Following an interval during which the growth rate of the cells gradually increases, the cells grow at a constant, maximum rate and this period is referred to as the *log* or *exponential phase*. As nutrient is exhausted, or toxic metabolites accumulate, the growth rate of the cells deviates from the maximum, and eventually growth ceases and the culture is said to enter the *stationary phase*. After a further period of time the viable cell number begins to decline and the culture enters the *death phase*.

The descriptive terminology of batch culture may be rather misleading when considering the metabolic activity of the culture during the various phases, for although the metabolism of stationary phase cells may be considerably different from that of logarithmic ones, it is by no means stationary. Bu'lock *et al.* (1965) proposed a descriptive terminology of the behaviour of microbial cells which considered the type of metabolism rather than the kinetics of growth. The term *trophophase* was used to describe the log, or exponential phase of a culture during which the sole products of metabolism are either essential to growth, such as amino acids, nucleotides, proteins, nucleic acids, lipids, carbohydrates etc., or the byproducts of energy-yielding catabolism, such as ethanol, acetone and butanol. The metabolites produced during the trophophase are described as primary metabolites. The term *idiophase* was used to describe the phase of a culture during which products other than primary metabolites are produced and these compounds are termed secondary metabolites. Secondary metabolites have been defined as compounds which are synthesized by slow-growing or non-growing cells, play no obvious role in cell growth and are taxonomically limited in their distribution. The idiophase, therefore, corresponds approximately to the stationary phase of batch culture.

The interrelationships between primary and secondary metabolism are illustrated in Fig. 6.1, from which it may be seen that secondary metabolites tend to be synthesized from the key intermediates and end-products of primary metabolism. Although the primary metabolic routes shown in Fig. 6.1 are common to the vast majority of micro-organisms, each secondary metabolite would be synthesized by only a very small number of microbial taxa. Also, not all microbial taxa undergo secondary metabolism; it is a common feature of the filamentous bacteria and fungi and the sporulating bacteria, but it is not, for example, a feature of the Enterobacteriaceae. Thus, although the taxonomic distribution of secondary metabolism is far more limited than that of primary metabolism, the range of different metabolites produced is enormous. At first sight it may seem anomalous that micro-organisms produce compounds that do not appear to have any metabolic function and are certainly not byproducts of catabolism. However, many secondary metabolites exhibit antimicrobial properties and, therefore, may be involved in competition in the natural environment (Demain, 1980); others, since their discovery in idiophase cultures, have been demonstrated to be produced during the trophophase where, it has been claimed, they may act in some regulatory role. Thus, some 'secondary metabolites' do not meet all the criteria quoted in the earlier definition, which should be considered more as a general description of the phenomenon rather than a true definition. Indeed, Campbell

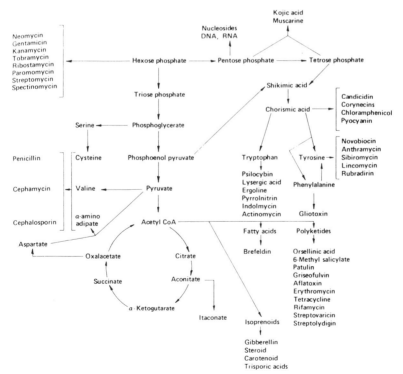

**Fig. 6.1** Interrelationships between primary and secondary metabolism. (Reproduced with permission from Malik, 1980)

(1984) claims that the only definitive aspect of this original definition is the taxonomically limited distribution of secondary metabolites. Although the inter-relationships between primary and secondary metabolism, and the physiological role of the latter group, are the subjects of considerable debate the relevance of these products to the fermentation industry is their commercial significance. Some examples of the commercial significance of such products are shown in Table 6.1.

The control of the onset of secondary metabolism has been the subject of extensive investigation and, obviously, this is an area of research of considerable interest to the fermentation industry. The outcome of this work is that there is a considerable amount of information available on the changes occurring in the medium at the onset of secondary metabolism but relatively little is known about the control of the process at the DNA level. Primary metabolic precursors of secondary metabolites have been demonstrated to induce their formation; for example, tryptophan in alkaloid bio-synthesis (Robbers and Floss, 1970) and methionine in cephalosporin synthesis (Komatsu, Mizumo and Kodaira, 1975). On the other hand, medium components have been demonstrated to repress secondary metabolism, the earliest observation being that of Soltero and Johnson in 1953 of the repressing effect of glucose on benzylpenicillin formation. Carbon, nitrogen and phosphate sources which support rapid growth have been variously shown to inhibit secondary metabolite production. Therefore, it is essential that repressing nutrients be avoided in fermentation media or be provided in sub-repressing levels; these aspects are considered in more detail in Chapter 8.

**Table 6.1** Some examples of microbial metabolites of commercial significance

| *Primary metabolites* | *Commercial significance* |
| --- | --- |
| Ethanol | 'Active ingredient' in alcoholic beverages |
| Citric acid | Various uses in the food industry |
| Acetone and butanol | Solvents |
| Glutamic acid | Flavour enhancer |
| Lysine | Feed additive |
| Polysaccharides | Applications in the food industry; enhanced oil recovery |
| $Fe^{+++}$ | Ore leaching |
| *Secondary metabolites* | |
| Penicillin | Antibiotic |
| Tetracyclines | Antibiotic |
| Cephalosporin | Antibiotic |
| Streptomycin | Antibiotic |
| Griseofulvin | Antifungal antibiotic |
| Pepstatin | Ulcer treatment |
| Cyclosporin A | Immunosuppressant |
| Krestin | Cancer treatment |
| Bestatin | Cancer treatment |

As mentioned in the introduction, the advent of recombinant DNA technology has extended the potential range of products that may be produced by micro-organisms. Microbial cells may be endowed with the ability to produce compounds normally associated with higher cells and such products may form the bases of new fermentation processes, for example the synthesis of interferon, insulin and renin. The techniques of *in vitro* recombinant DNA technology are considered in Chapters 10, 11 and 12.

The biosynthetic and catabolic pathways occurring during primary and secondary biosynthesis are all catalyzed by enzymes. As illustrated in Table 6.2, many microbial enzymes have commercial application and are, therefore, also worthwhile commercial fermentation products. Most wild-type micro-organisms regulate the synthesis of metabolic products such that they are produced only in the amounts necessary to meet the growth requirements of the cells. However, in a commercial process it is essential to produce these compounds at very high concentrations and, therefore necessary to modify the producing cells and to control their growth environment such that their normal, limiting, control systems are overcome. The control of the organism's environment and its genetic modification are considered in Chapters 8 and 9.

## 6.4 Microbial transformation processes

As well as the use of micro-organisms to produce biomass and microbial metabolites, microbial cells may be used to catalyse the conversion of a compound into a structurally similar, financially more valuable, compound. Although the production of vinegar is the oldest and most well-established transformation process (the conversion of ethanol into acetic acid) the majority of these processes involve the production of high value compounds. The reactions which may be catalysed include oxidation, dehydrogenation, hydroxylation, dehydration, condensation, decarboxylation, amination, deamination and isomerization. Microbial processes have the advantage of specificity over the use of chemical reagents and of operating at relatively low temperatures and pressures. The anomaly of transformation processes is that a large biomass has to be produced to catalyse, perhaps, a single reaction. Thus, some

**Table 6.2** Some examples of the commercial applications of enzymes

| Enzyme | Microbial source | Industry | Application |
|---|---|---|---|
| Amylase | Fungal | Baking | Fermentation acceleration, reduction of dough viscosity |
| Amylase | Fungal/bacterial | Brewing | Release of soluble sugars in mashing |
| Protease | Fungal/bacterial | Brewing | Protection against 'chill haze.' |
| Amylase | Fungal/bacterial | Food | Manufacture of syrups |
| Pectinase | Fungal | Food | Preparation of fruit juices and coffee concentrates |
| Protease | Bacterial | Laundry | Enzyme incorporated into washing powders |

processes have been streamlined by immobilizing either the cells themselves or the isolated enzymes which catalyze the reaction, on an inert support which may then be utilized as reusable catalysts. This aspect is considered in more detail in Chapter 14.

## 6.5   The structure of a fermentation process

The central part of a fermentation process is the growth of the industrial organism in an environment which stimulates the synthesis of the desired commercial product. This is carried out in a fermenter which is, essentially, a large vessel (ranging in size from 1000 to 1.5 million dm$^3$) in which the organism may be maintained at the required temperature, pH, dissolved oxygen concentration and substrate concentration. However, the actual culturing of the organism in the fermenter is only one of a number of stages in a fermentation process, as indicated in Fig. 6.2. The medium on which the organism is to be grown has to be formulated from its raw materials and sterilized; the fermenter has to be sterilized and inoculated with a viable, metabolically active culture which is capable of producing the required product; after growth, the culture fluid has to be harvested, the cells separated from the supernatant and the product has to be extracted from the relevant fraction (either the cells or the cell-free supernatant) and purified. One must also visualize the research and development programme superimposed upon this process. For a fermentation to be possible in the first place, an acceptable, productive organism must be obtained (normally by screening natural isolates) and its productivity raised to economic levels by medium improvement, mutation, recombination and process design. Thus, a successful fermentation is based on the skills of microbiologists, biochemists, geneticists, chemical engineers, chemists and control engineers. Although this text concentrates on the biological aspects of biotechnology it is essential not to lose sight of the interdisciplinary nature of the subject.

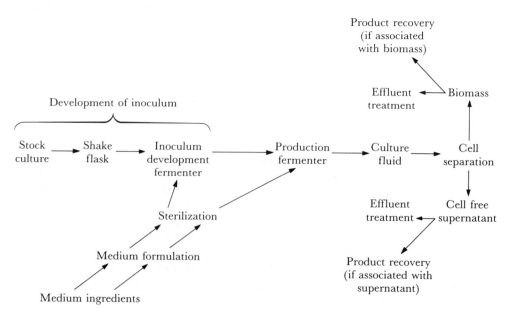

**Fig. 6.2**   The stages in a fermentation process

# Chapter 7

# *Culturing micro-organisms*

## 7.1  Batch culture

A micro-organism will grow in a culture medium provided that the medium contains all the necessary nutrients in an available form and that all other environmental factors are suitable. The simplest culture method is batch culture in which the micro-organism is grown on a limited amount of medium until either one essential nutrient component is exhausted or toxic byproducts accumulate to growth-inhibiting levels. In a batch system the culture will pass through a number of stages, as illustrated in Fig. 7.1.

Following inoculation of the culture medium growth does not occur immediately and the period prior to active growth is referred to as the lag phase and may be considered as a period of adaptation. In a commercial process it is obviously desirable to reduce the lag phase as much as possible, for not only is time wasted but also the medium is consumed in maintaining the culture viable prior to growth. The length of the lag phase may be reduced by using an exponentially growing, relatively large inoculum (3–10%, by volume) which has itself been grown in the same (or similar) medium to be used in the fermentation.

### 7.1.1  *Inoculum development*
The culture used to inoculate a plant fermenter must be available in sufficient quantity, be in a metabolically active state, free from contaminants and capable of producing the desired product in subsequent culture. However, if, for example, the production fermenter were 100 000 dm$^3$ then the volume of the inoculum should be between 3000 and 10 000 dm$^3$. This volume of culture would have to be built up from an initial stock culture of a few cm$^3$ which would involve a large number of successive seed fermentations of progressively increasing scale. However, there is a conflict between using the 'optimum' inoculum volume and producing an inoculum that is pure and free from degenerate strains (that is, cells which have lost the ability to produce the desired product), because the greater the number of stages between the stock culture and the final fermentation the greater is the risk of contamination and strain degeneration. Also, very considerable capital outlay would be required to

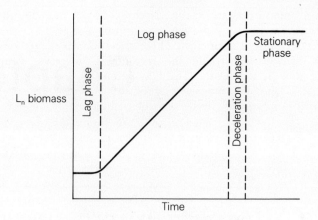

**Fig. 7.1** Schematic representation of the growth of a micro-organism in batch culture

provide the inoculum development fermenters. If the fermentation plant contains a large number of production scale fermenters, one series of inoculum development vessels may be adequate to supply all the inoculum needs, which would reflect reasonable return on investment. However, if the production plant consisted of just one very large continuous culture vessel then it would hardly be economic to construct an inoculum fermenter, one-tenth of the capacity of the production vessel, which would be used only infrequently. Thus, a compromise is reached between the size of the inoculum used, the probabilities of contamination and strain degeneration, and the investment costs involved.

Many industrially important fermentation organisms are filamentous and are capable of producing asexual spores. Such organisms include *Penicillium chrysogenum* (used for commercial penicillin production), *Aspergillus niger* (used for the commercial production of organic acids) and some of the actinomycetes (used for the commercial production of a range of antibotics and other products).

The property of asexual spore production is extremely useful in the development of inoculum because a relatively small biomass may be used to produce a very high concentration of microbial propagules. Stanbury and Whitaker (1984) discuss the use of spore inocula in detail, but the major points will be summarized here. Mycelial organisms will sporulate profusely on agar-solidified media, or on the surface of cereal grains, and both culture methods are used for the production of spore inocula for industrial purposes (Vezina and Singh, 1975). Butterworth (1984) described the use of a Roux bottle giving approximately 200 cm$^2$ of agar surface for the sporulation of *Streptomyces clavuligerus* (used for the commercial production of clavulanic acid, an inhibitor of $\beta$-lactamase). The spores produced from one such vessel could be used to inoculate a 75 dm$^3$ seed fermentation, which was subsequently used to inoculate a 1500 dm$^3$ production fermenter. Podojil *et al.* (1984) cite the use of flasks of millet for the production of spores of *Streptomyces aureofaciens* in the commercial production of chlortetracycline. Many fungi will sporulate in submerged culture (that is, in a stirred fermenter) provided that the optimum medium is utilized (Vezina, Singh and Sehgal, 1965); for example, *Penicillium patulum* spores are produced in submerged culture in the inoculum development programme for the commercial production of griseofulvin (Huber and Tietz, 1984).

The filamentous nature of many commercially important organisms does present considerable problems to the fermentation technologist. A filamentous organism may give rise to various growth forms in submerged liquid culture, only some of which may be desirable for the commercial process. These growth forms range from long filaments homogeneously distributed through the medium to discrete, compact pellets of mycelium suspended in the broth. The filamentous type of habit gives rise to an extremely viscous broth which may be very difficult to aerate adequately, whereas the pellet type of habit gives rise to a far less viscous, but also less homogeneous, broth. In a pelleted culture the mycelium at the centre of the pellets may be starved of nutrients and oxygen due to diffusion limitations. Also, there is considerable evidence that the morphological form of the organism influences the productivity of the culture, but whether this is due to the phenomena already mentioned or to some form of metabolic control is far from clear (Whitaker and Long, 1973; Calam and Smith, 1981). Thus, some fermentations are carried out with the organism in a filamentous habit whereas others are carried out with the organism growing as pellets. For example, filamentous growth has been claimed to be optimum for penicillin production from *P. chrysogenum* whereas pelleted growth has been claimed to be optimum for citric acid production from *A. niger* (Smith and Calam, 1980; Al Obaidi and Berry, 1980). The necessity for filamentous growth is taken to the extreme in the ICI-RHM mycoprotein process where *Fusarium graminearium* is produced for human consumption. A highly filamentous morphology is required to produce the desired texture in the product which resembles the strength and eating texture of white and soft red meats (Marsh and Pinkney, 1985). Thus, in this process a median hyphal length of 400 $\mu$m is required.

The relevance of this consideration of mycelial morphology to inoculum development is that the morphology may be influenced considerably by the concentration of spores used as inoculum for the fermentation. A high spore concentration tends to give rise to a filamentous habit, whereas a relatively low spore concentration tends to give rise to a pelleted one (Whitaker and Long, 1973). Also, the nature of the medium used in the inoculum development programme will influence the morphological form. Although the vast majority of the information in the literature relates to the morphology of fungi in submerged culture, the subject is probably equally important in the actinomycete fermentations.

Figure 7.2 includes some examples of commercial inoculum development programmes, from which it may be seen that the nature of the programme varies quite considerably depending on the process and it must be remembered that the success of an inoculum development system is judged by the productivity of the developed culture in the subsequent fermentation. From the above account it should be obvious that it is extremely undesirable to use cells from one production fermentation to inoculate the next one; the procedure is that an inoculum is built up from the original stock culture for each fermentation run. However, there are two exceptions to this rule and these are the brewing of beer and of vinegar. In the production of vinegar by batch fermentation using acetic acid bacteria cells from the end of a fermentation are used to inoculate the next one by withdrawing approximately 60% of the culture and restoring the original volume with fresh medium. The production organism is extremely sensitive to oxygen limitation and it is crucial to have a rapidly growing culture; also, the risks of contamination and strain degeneration are less due to the highly selective environment of the vinegar fermentation. In traditional ale breweries it has been common practice to inoculate (or pitch, in brewing terms) a fermentation with yeast obtained from a previous fermentation, and this procedure may continue for many years. However, as early as 1896 Hansen developed a system for the production of pure inocula for the brewing of lager. In most modern ale breweries inoculation from brew to brew is now relatively limited but it is still practised in many of the small, traditional breweries and it should be added that the quality of their product is usually exceptional!

### 7.1.2 *Logarithmic or exponential growth*

Provided that an active inoculum is used (as described above) the length of the lag phase may be minimized and the growth rate of the cells gradually increases. Eventually, the cells grow at a constant, maximum rate and this period is referred to

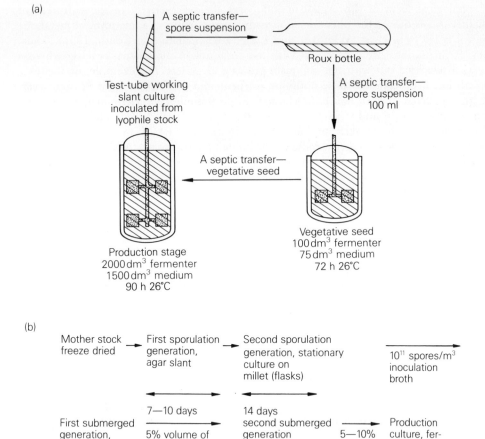

(a)

A septic transfer— spore suspension

Roux bottle

Test-tube working slant culture inoculated from lyophile stock

A septic transfer— spore suspension 100 ml

A septic transfer— vegetative seed

Production stage 2000 dm³ fermenter 1500 dm³ medium 90 h 26°C

Vegetative seed 100 dm³ fermenter 75 dm³ medium 72 h 26°C

(b)

| Mother stock freeze dried | → | First sporulation generation, agar slant | → | Second sporulation generation, stationary culture on millet (flasks) | | 10¹¹ spores/m³ inoculation broth |

Mother stock freeze dried → First sporulation generation, agar slant → Second sporulation generation, stationary culture on millet (flasks) → $10^{11}$ spores/m³ inoculation broth

7—10 days      14 days

First submerged generation, preinoculation tank → 5% volume of preinoculation tank      second submerged generation inoculation tank → 5—10% volume of inoculation tank → Production culture, fermentation tank

24—26 h          18—20 h          100—200 h

**Fig. 7.2** Some representative examples of commercial inoculum development programmes. (a) Clavulanic acid production by *Streptomyces clavuligerus*. (Reproduced with permission from Butterworth, 1984.) (b) Chlortetracycline production by *Streptomyces aureofaciens*. (Reproduced with permission from Podojil *et al.*, 1984.) (c) Sagamycin production by *Streptomyces sp*. (Reproduced with permission from Okachi and Nara, 1984)

(c)

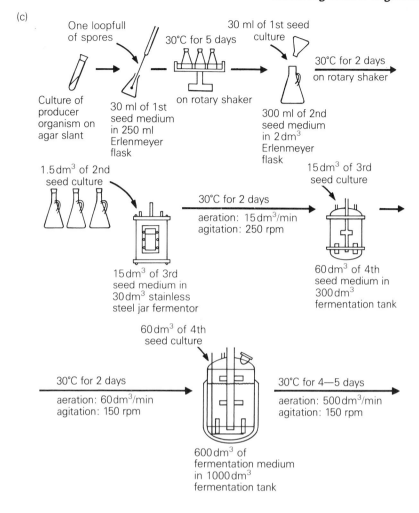

One loopfull of spores

30°C for 5 days

30 ml of 1st seed culture

30°C for 2 days on rotary shaker

Culture of producer organism on agar slant

30 ml of 1st seed medium in 250 ml Erlenmeyer flask

on rotary shaker

300 ml of 2nd seed medium in 2 dm³ Erlenmeyer flask

1.5 dm³ of 2nd seed culture

30°C for 2 days

aeration: 15 dm³/min
agitation: 250 rpm

15 dm³ of 3rd seed culture

15 dm³ of 3rd seed medium in 30 dm³ stainless steel jar fermentor

60 dm³ of 4th seed medium in 300 dm³ fermentation tank

60 dm³ of 4th seed culture

30°C for 2 days

aeration: 60 dm³/min
agitation: 150 rpm

30°C for 4—5 days

aeration: 500 dm³/min
agitation: 150 rpm

600 dm³ of fermentation medium in 1000 dm³ fermentation tank

**Fig.** 7.2 (*Continued*)

as the log, or exponential phase which may be described by the equation:

$$\frac{dx}{dt} = \mu x \tag{1}$$

where $x$ is the cell concentration (mg cm$^{-3}$)
$t$ is the time of incubation (h)
$\mu$ is the specific growth rate (h$^{-1}$).

On integration, equation (1) gives:

$$x_t = x_0 e^{\mu t} \tag{2}$$

where $x_0$ is the cell concentration at time 0
$x_t$ is the cell concentration after a time interval of $t$ h.

On taking natural logarithms of equation (2) we obtain:

$$\ln x_t = \ln x_0 + \mu t.$$

Thus, a plot of the natural logarithm of the cell concentration against time gives a straight line, the slope of which equals $\mu$, the specific growth rate. During the exponential phase the specific growth rate is constant, and maximum for the operating conditions and is thus termed $\mu_{max}$, the maximum specific growth rate.

The growth of an organism in batch culture may also be described by simple algebra using the term doubling time, $t_d$, which is the time taken for a cell to divide. If a culture medium were inoculated with one viable cell, and this cell was capable of growing at the maximum rate without any intervening lag phase, then the sequential increase in cell number that would be observed would be: 1, 2, 4, 8, 16, 32, 64, 128, etc. This sequential increase may be represented by expressing the cell numbers to the base two, thus: $2^1$, $2^2$, $2^3$, $2^4$, $2^5$, $2^6$, $2^7$, $2^8$, etc. The exponents (or power numbers or indices) in this series represent the number of generations which have past, thus after $n$ generations there will be $2^n$ cells in the culture. If the medium were inoculated with $N_0$ cells then the number of cells, $N_t$, present after $n$ generations would be:

$$N_t = N_0 \cdot 2^n$$

or

$$N_t = N_0 \cdot 2^{t/t_d}. \qquad (3)$$

On taking logarithms to the base 2, equation (3) becomes:

$$\mathrm{Log}_2 \, N_t = \mathrm{Log}_2 \, N_0 + \frac{t}{t_d} \qquad (4)$$

Thus

$$t_d = \frac{t}{\mathrm{Log}_2 \, N_t - \mathrm{Log}_2 \, N_0}. \qquad (5)$$

Although the term doubling time is relatively little used in microbial physiology it may be useful to convert values for specific growth rate to doubling time to obtain a better appreciation of the meaning of these values. The equation derived for doubling time, using number of cells may also be derived using cell concentration, thus:

$$t_d = \frac{t}{\mathrm{Log}_2 \, x_t - \mathrm{Log}_2 \, x_0}. \qquad (6)$$

It has already been shown that:

$$\ln x_t = \ln x_0 + \mu t.$$

If, in this equation $t = td$, then $x_t = 2x_0$:

$$\ln 2x_0 = \ln x_0 + \mu t_d$$

or

$$\mu = \frac{\ln 2}{t_d} \quad \text{and} \quad t_d = \frac{\ln 2}{\mu}.$$

Therefore

$$\mu = \frac{0.693}{t_d}. \tag{7}$$

Thus, a specific growth rate of 1 h$^{-1}$ is equivalent to a doubling time of 0.693 h. Table 7.1 gives some representative values of $\mu_{max}$ for a range of micro-organisms.

Equations (2) and (3) ignore the facts that growth results in the depletion of nutrients and the accumulation of toxic byproducts and thus predict that growth continues indefinitely. However, as substrate is exhausted or toxic byproducts accumulate the specific growth rate of the cells deviates from the maximum, growth eventually ceases and the culture enters the stationary phase.

The decrease in growth rate and the cessation of growth due to the depletion of substrate may be described by the relationship between $\mu$ and the residual limiting substrate concentration. The limiting substrate is that component of the medium which is the first to become exhausted. Monod (1942) demonstrated the following relationship:

$$\mu = \frac{\mu_{max} s}{K_s + s} \tag{8}$$

where $s$ is the residual concentration of the limiting substrate, and $K_s$ is the utilization or saturation constant for the limiting substrate and is equivalent to the substrate concentration when $\mu$ is half $\mu_{max}$.

Equation (8) is represented graphically in Fig. 7.3 where zone B to C is equivalent to the exponential phase of batch culture with substrate in excess and growth at $\mu_{max}$. The zone A to B is equivalent to the deceleration in the growth rate from the end of the exponential phase to the onset of the stationary phase. The numerical value of $K_s$ reflects the organism's affinity for its substrate—a high $K_s$ value indicates a low affinity and a low $K_s$ value indicates a high affinity. Thus, if the organism has a very high affinity for the limiting substrate (low $K_s$ value) the specific growth rate in batch culture will not be affected until the substrate concentration has declined to a very low level. Thus, such an organism would display only a very short deceleration phase. Likewise, if an organism has a very low affinity for the limiting substrate (a high $K_s$

**Table 7.1** Representative values of $\mu_{max}$ for a range of micro-organisms (obtained under the conditions specified in the references)

| Organism | $\mu_{max}$ (h$^{-1}$) | Reference |
| --- | --- | --- |
| *Beneckea natriegens* | 4.24 | Eagon, 1961 |
| *Methylomonas methanolytica* | 0.53 | Dostalek, Haggstrom and Molin, 1972 |
| *Penicillium chrysogenum* | 0.12 | Trinci, 1969 |

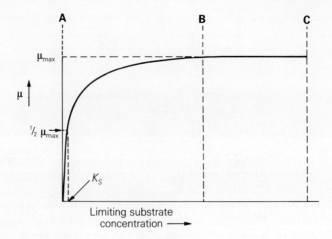

**Fig. 7.3**   The effect of residual limiting substrate concentration on the specific growth rate of a micro-organism

value) the specific growth rate will decrease at a relatively high substrate concentration and, therefore, such an organism would display a long deceleration phase. Some representative values for $K_s$ of a range of organisms and substrates are given in Table 7.2.

The amount of biomass at the stationary phase is dependent on the composition of the medium and the efficiency of the organism in converting the substrates into cells. Ideally, the medium should be designed such that growth is limited by the exhaustion of a single substrate rather than by the accumulation of toxin. The concentration of cells at the stationary phase is then given by the equation:

$$x = Y \cdot S_R \tag{9}$$

where $x$ is the cell concentration,

$Y$ is the yield factor for the limiting substrate (g of biomass per g of substrate consumed),

$S_R$ is the original substrate concentration in the medium.

**Table 7.2**   Some representative values of $K_s$ for a range of organisms and substrates

| Organism | Substrate | $K_s$ (mg dm$^{-3}$) | Reference |
|----------|-----------|----------------------|-----------|
| *Escherichia coli* | Glucose | $6.8 \times 10^{-2}$ | Shehata and Marr (1971) |
| *Escherichia coli* | Phosphate ions | 1.6 | Shehata and Marr (1971) |
| *Pseudomonas* sp. | Methanol | 0.7 | Harrison (1973) |
| *Aspergillus niger* | Glucose | 5.0 | Pirt (1973) |
| *Saccharomyces cerevisiae* | Glucose | 25 | Pirt and Kurowski (1970) |

The term $Y$ is a measure of the cell efficiency in converting substrate into biomass. The biomass at any point in the batch culture is given by the equation:

$$x = Y(S_R - s)$$

where $s$ is the residual substrate concentration at that point in time.

Thus,

$$Y = \frac{x}{(S_R - s)}$$

therefore, provided that cell and substrate concentration are in the same units, $Y$ is a dimensionless constant. The value of $Y$ will obviously depend on organism and substrate and the importance of this value is discussed in more detail in 8.3.1.


## 7.2 Continuous culture

If a culture medium is designed such that the cessation of growth is due to the exhaustion of a medium component (that is, substrate limited) rather than by the accumulation of toxins then exponential growth in batch culture may be prolonged by the addition of fresh medium to the culture vessel. This procedure may be repeated several times until the vessel is full. However, if an overflow were to be installed in the side of the fermenter such that the addition of fresh medium displaced an equal volume of culture then continuous production of cells could be achieved. If medium were added continuously to such a system at a suitable rate the displacement of culture may be balanced by the production of new biomass, and thus a steady state may be achieved. The growth of the cells in a continuous culture of this type is controlled by the availability of the growth-limiting chemical component in the medium, and the system is described as a chemostat. It is important to appreciate that any component of the medium may be made the growth-limiting nutrient, and that the nature of the limitation will affect markedly the physiology and biochemistry of the cells.

### 7.2.1 *The kinetics of continuous culture*
The kinetics of continuous culture may be described as follows:

The flow of medium through the system is represented by the term dilution rate, $D$, defined as

$$D = \frac{F}{V}$$

where $D$ is the dilution rate $(h^{-1})$
$V$ is the volume $(dm^3)$
$F$ is the flow rate $(dm^3 h^{-1})$.

The change in the cell concentration in the fermenter over a time period may be described as:

$$\frac{dx}{dt} = \text{growth} - \text{output}$$

or

$$\frac{dx}{dt} = \mu x - Dx. \tag{10}$$

As suggested earlier the system enters a steady state where the output of biomass is compensated for by the growth of the cells, thus:

$$\frac{dx}{dt} = 0$$

$$\mu x = Dx$$

$$\mu = D. \tag{11}$$

Equation (11) states that under steady state conditions in a chemostat the specific growth rate is controlled by the dilution rate and is thus under the control of the operator of the experiment. The mechanism underlying the controlling effect of the dilution rate is, essentially, the relationship between $\mu$ and $s$ as demonstrated by Monod (1942):

$$\mu = \frac{\mu_{max} s}{K_s + s}.$$

At steady state $\mu = D$ and therefore:

$$D = \frac{\mu_{max} \bar{s}}{K_s + \bar{s}} \tag{12}$$

where $\bar{s}$ is the steady state residual substrate concentration in the medium

Rearrangement of equation (12) gives:

$$\bar{s} = \frac{K_s D}{\mu_{max} - D}. \tag{13}$$

Equation (13) predicts the substrate concentration in the chemostat is determined by the dilution rate. In effect, this occurs by growth depleting the substrate to a concentration that supports the growth rate equal to the dilution rate.

If the substrate is depleted below the level that supports the growth rate dictated by the dilution rate the following sequence of events takes place:

(1) The growth rate of the cells will be less than the dilution rate, and they will be washed out of the vessel at a rate greater than they are being produced resulting in a decrease in biomass.
(2) The substrate concentration in the vessel will rise because fewer cells are left in the vessel to consume the substrate.
(3) The increased substrate concentration will result in the cells growing at a rate greater than the dilution rate and biomass concentration will increase.

The concentration of cells in the chemostat at steady state is described by the equation:

$$\bar{x} = Y(S_R - \bar{s}) \tag{14}$$

where $\bar{x}$ is the steady state cell concentration.

By combining equations (13) and (14), then:

$$\bar{x} = Y\left(S_R - \frac{K_s D}{\mu_{max} - D}\right). \tag{15}$$

Thus, the biomass concentration at steady state is determined by the operational variables, $S_R$ and $D$. If $S_R$ is increased, $\bar{x}$ will increase but $\bar{s}$, the residual substrate concentration in the chemostat, will remain the same. If $D$ is increased, $\mu$ will increase ($\mu = D$) and the residual substrate would have increased to support the elevated growth rate; thus, less substrate will be available to be converted into biomass which will then decrease.

The effects of increasing dilution rate and initial substrate concentration are illustrated in Fig. 7.4. From Fig. 7.4 it may be seen that as $D$ increases, $\bar{s}$ increases only very slightly. This is due to the fact that the $K_s$ value is usually very small and, thus, a significant increase in dilution rate (specific growth rate) may be supported by only a slight increase in residual substrate concentration. As $D$ approaches $\mu_{max}$ the concentration of residual limiting substrate required to support the increased growth rate increases significantly, and thus residual substrate rises and biomass declines. Eventually, a dilution rate is reached at which $\bar{x}$ equals zero (that is, all cells have been washed out of the vessel and $\bar{s} = S_R$. This dilution rate is referred to as $D_{crit}$ and is given by the equation:

$$D_{crit} = \frac{\mu_{max} \cdot S_R}{K_s + S_R}.$$

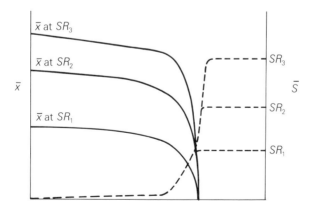

**Fig. 7.4** The effect of dilution rate and increasing initial substrate concentration (SR3 > SR2 > SR1) on biomass and residual limiting substrate concentration in a steady state chemostat; dilution rate: ——— biomass ($\bar{x}$) concentration; – – – – residual limiting substrate concentration

$D_{crit}$ is always slightly lower than $\mu_{max}$ as $\mu_{max}$ may never be achieved in a simple chemostat because substrate-limiting conditions must always prevail.

Some organisms display a very poor affinity for substrate (a high $K_s$ value) and, thus, as $D$ is increased so residual substrate concentration must increase significantly to support the larger growth rate. Thus, for such an organism, $\bar{x}$ shows a significant decrease over a relatively small range of dilution rates.

### 7.2.2   *Differences between batch and continuous culture*
The major differences between batch and continuous culture may be summarized as follows:

(1) Continuous culture operates under steady state conditions (the concentrations of all components in the culture are constant).
(2) Continuous culture operates under substrate limiting conditions whereas a batch culture, in the exponential phase, operates in the presence of excess substrate.
(3) The growth rate in continuous culture is controlled by the dilution rate (and is therefore under the control of the operator) and is always less than $\mu_{max}$, whereas the specific growth rate in the exponential phase of batch culture will be at the $\mu_{max}$ for the prevailing culture conditions.

## 7.3   Fed-batch culture

The term fed-batch culture is used to describe batch cultures which are fed, continuously or sequentially, with fresh medium without the removal of culture fluid. Thus, the volume of a fed-batch culture increases with time. Pirt (1975) described the kinetics of such a system as follows:

If the growth of an organism in batch culture were limited by the concentration of one substrate in the medium, the biomass concentration at stationary phase, $x_{max}$, would be described by the equation:

$$x_{max} \simeq Y \cdot S_R$$

assuming that the initial inoculum was insignificant compared with the final biomass. If fresh medium were to be added to the vessel at a dilution rate less than $\mu_{max}$ then virtually all the substrate would be consumed as it entered the system:

$$F \cdot S_R \simeq \mu \cdot \frac{X}{Y}$$

where $X$ is the total biomass present in the vessel ($Vx$). Although the total biomass ($X$) in the vessel increases with time the concentration of cells, $x$, remains virtually constant, that is

$$\frac{dx}{dt} \simeq 0$$

and, therefore, $\mu = D$.

Such a system is then said to be in quasi-steady state. As time progresses and the volume of the culture increases, so the dilution rate will decrease. Thus, the value of $D$ is given by the expression:

$$D = \frac{F}{V_0 + F \cdot t}$$

where $F$ is the flow rate,
  $V_0$ is the initial volume of the culture and
  $t$ is the time that fed-batch conditions have been operating.

Monod kinetics predict that as $D$ falls the residual substrate concentration should also decrease, resulting in an increase in the biomass concentration. However, over the range of growth rates operating the initial substrate concentration will be much larger than the residual concentration and the increase in $\bar{s}$ will be insignificant. The major difference between the steady state of a chemostat and the quasi-steady state of a fed-batch culture is that in a chemostat the dilution rate is constant, whereas in a fed-batch system $D$ decreases with time. The dilution rate in a fed-batch culture may be kept constant by increasing exponentially the flow rate, using a computer control system.

## 7.4 The use of culture systems for the production of microbial products

The self-balancing properties and the steady state condition of the chemostat, theoretically, make it a very attractive culture system for the production of microbial products. However, for a variety of reasons, the large-scale industrial use of continuous culture is extremely limited and is confined to the production of biomass and ethanol, not that it is true to say that the predominant system currently in use is batch culture. Most industrial batch processes have, to a greater or lesser extent, been modified to operate as fed-batch systems resulting in the achievement of some of the advantages of continuous culture without the major drawbacks.

### 7.4.1 *Comparison of batch and continuous culture as industrial production systems*
The productivity of a culture system, in terms of cells produced, may be described as the output of biomass per unit time. Stanbury and Whitaker (1984) described the productivity of a batch culture as:

$$R_{batch} = \frac{x_{max} - x_0}{t_i + t_{ii}} \tag{16}$$

where $R_{batch}$ is the output of biomass $(g\ dm^{-3}\ h^{-1})$ of the batch culture.
  $x_{max}$ is the concentration of biomass at stationary phase $(g\ dm^{-3})$
  $x_0$ is the inoculum biomass concentration $(g\ dm^{-3})$
  $t_i$ is the time during which the organism grows at $\mu_{max}$
  $t_{ii}$ is the time during which the organism is not growing at $\mu_{max}$ and includes the lag phase, the deceleration phase and the periods of batching, sterilizing and harvesting;

and the productivity of a continuous culture as:

$$R_{\text{cont}} = D\bar{x}\left(1 - \frac{t_{\text{iii}}}{T}\right) \qquad (17)$$

where $R_{\text{cont}}$ is the output of biomass $(\text{g dm}^{-3}\,\text{h}^{-1})$ of the continuous culture

$t_{\text{iii}}$ is the time period prior to the establishment of a steady state and includes time for vessel preparation, sterilization and operation in batch culture prior to continuous operation.

$T$ is the time period during which steady state conditions prevail.

The term $D\bar{x}$, in equation (17), increases to a maximum with increasing dilution rate after which any further increase in $D$ results in a decrease in $D\bar{x}$ until $D_{\text{crit}}$ is reached, at which point output is zero. Therefore, maximum output of biomass per unit time (that is, productivity) may be achieved by operating at the dilution rate giving the highest value of $D\bar{x}$, this value being referred to as $D_{\text{max}}$. The batch fermentation described in equation (16) is an average for the total time of the fermentation. Because $dx/dt = \mu x$ the productivity of the culture increases with time, and thus the vast majority of the biomass is produced towards the end of the exponential phase of the fermentation. In a steady state chemostat operating at, or near to, $D_{\text{max}}$ the productivity remains constant, and maximum for the whole of the fermentation. Also, a continuous process may be operated for a very long time so that the non-productive period, $t_{\text{iii}}$, in equation (17) may be insignificant. However, the non-productive time element $(t_{\text{ii}})$ for a batch culture is a very significant period of time, especially as a batch culture would have to be re-established many times during the running time of a comparable continuous process. Thus, $t_{\text{ii}}$ would be a recurrent factor, so a continuous process is obviously far more productive for the manufacture of biomass than a comparable batch one.

The steady state nature of the continuous process is also advantageous in that the system should be far easier to control than a comparable batch one. During a batch culture heat output, acid or alkali production and oxygen consumption will range from very low rates at the start of the fermentation to very high rates during late logarithmic growth. Thus, the control of the environment of such a system is far more difficult than that of a continuous process where, at steady state, production and consumption rates should be constant. Furthermore, a continuous process should result in a more constant labour demand than a comparable batch one.

A frequently quoted disadvantage of continuous systems is their susceptibility to contamination by 'foreign' organisms. The control of contamination is essentially a problem of fermenter design, construction and operation and should be overcome by good engineering and microbiological practice. Smith (1980) discussed the approach used in the design and operation of the ICI Pruteen continuous fermenter (working volume $15\,000\,\text{m}^3$), from which it is obvious that it is perfectly feasible to build and operate a very large-scale chemostat. Thus, problems of contamination should not be overemphasized. (Some aspects of the design of aseptic fermentations are considered in 8.2.2.)

The superiority of continuous culture for the production of biomass has been recognized and has come to fruition in the ICI process. The production of growth-associated byproducts should also be more efficient in continuous culture. However, the success of such systems has been confined to the production of commercial ethanol whereas the production of potable alcohol by continuous culture has met with only limited success (at least in the United Kingdom).

The brewing industry in the United Kingdom has used continuous culture for the production of beer on a commercial scale but further improvements in the batch system have subsequently rendered the continuous process redundant. It must be remembered that it is not the brewer's objective to simply produce alcohol—he must produce a palatable drink acceptable to a 'discerning' clientele. Thus, beer is not fit for consumption immediately after the fermentation but must be stored, under suitable conditions, for up to 2 weeks after production for 'conditioning' to take place. Continuous brewing reduces the fermentation time for a beer from about 1 week for a traditional batch-process to 4–8 h. However, the adoption of the cylindroconical batch vessel (described by Nathan in 1930 but not adopted until the 1970s) resulted in a batch fermentation being reduced to approximately 48 h. Although this is still significantly longer than the continuous system when the total time for production is considered (that is, the fermentation plus conditioning time) the speed of the continuous system does not outweigh its disadvantages, which are:

(1) the long-term start-up period for the system to establish a steady state;
(2) the inflexibility of the system—it is very time-consuming to change a fermenter from one type of beer to another;
(3) the difficulty in matching the flavour of the continuous product to that of the traditional batch product.

Thus, continuous brewing has been almost totally superseded by improved batch systems.

The adoption of continuous culture for the production of microbial products has been extremely limited. Although, theoretically, it is possible to optimize a continuous process such that maximum productivity of a metabolite should be achieved, the long-term stability of such systems is precarious due to the problem of strain stability. A consideration of the kinetics of continuous culture reveals that the system is highly selective and will favour the propagation of the best-adapted organism in a culture. 'Best-adapted' in this context refers to the organism's affinity for the limiting substrate at the operating dilution rate. Thus, if organism $P$ is capable of maintaining the specific growth rate dictated by the dilution rate at a lower residual limiting substrate concentration than organism $Q$, then organism $P$ will predominate in the culture, and organism $Q$ will eventually be eliminated from the chemostat. A commercial organism is usually highly mutated and may contain genetic elements introduced by *in vitro* genetic manipulation such that it will produce very high concentrations of the desired product. Therefore, in physiological terms, such commercial organisms are extremely inefficient and a revertant strain, producing less of the desired product, may be better adapted to the cultural conditions than the superior producer, and will dominate the culture. This phenomenon, termed by Calcott (1981) 'contamination from within', is the major reason for the lack of use of continuous culture for the production of microbial metabolites.

### 7.4.2 *Fed-batch culture for the commercial production of metabolites*

As already discussed, although industry has been reluctant to adopt continuous culture for the production of microbial metabolites, major advances have been made in the development of fed-batch systems. Fed-batch culture may be used to extend the productive period of a traditional batch process without the inherent disadvantages of a continuous process and provide the fermentation technologist with a means of controlling the fermentation. The use of fed-batch systems for the control of fermentations is discussed in Section 8.4.

## 7.5   Summary

Thus, from this account it may be seen that micro-organisms may be grown in batch, fed-batch and continuous culture. Continuous culture is a system that provides the microbiologist with a means of maintaining a culture in a steady state environment in a controlled physiological condition. However, although continuous culture is an excellent research tool its use in industrial fermentation is confined to the production of biomass. The most widely adopted industrial scale culture method is fed-batch culture which allows the fermentation technologist considerable process control without the inherent difficulties presented by continuous culture systems.

# Chapter 8

## The control of the environment of the process organism

### 8.1  Introduction

The environment provided for the growth of the process organism must be controlled during the fermentation such that maximum (and reliable) productivity may be achieved. The important environmental factors are:

(1) The chemical environment of the organism should be such that it supports optimum product formation commensurate with the economics of the process.
(2) The temperature of the process should be optimum for product formation.
(3) The culture should be maintained in a pure state throughout the fermentation.

These environmental factors are influenced by the following:

(1) the design of the fermenter and the conditions under which it is operated;
(2) the design of the culture medium;
(3) the mode of operation of the fermentation.

### 8.2  Fermenter design and operation

A detailed consideration of fermenter design is outside the scope of this book which concentrates on the biological principles of biotechnology. However, fermentation technology is an amalgam of biology and chemical engineering and, therefore, it is necessary to give a brief summary of the types of fermenter available and their major design features. Stanbury and Whitaker (1984) listed the following 13 points considered to be important criteria in the design of a fermenter:

(1) The vessel should be capable of being operated aseptically for a number of days and should be reliable in long-term operation.

(2)  Adequate aeration and agitation should be provided to meet the metabolic requirements of the micro-organism.
(3)  Power consumption should be as low as possible.
(4)  A system of pH control should be provided.
(5)  Sampling facilities should be provided.
(6)  A system of temperature control should be provided.
(7)  Evaporation losses from the fermenter should not be excessive.
(8)  The vessel should be designed to require the minimal use of labour in operation, harvesting, cleaning and maintenance.
(9)  The vessel should be suitable for a range of processes.
(10)  The vessel should be constructed to ensure smooth internal surfaces, using welds instead of flange joints wherever possible.
(11)  The vessel should be of similar geometry to both smaller and larger vessels in the plant or pilot plant to facilitate scale-up.
(12)  The cheapest materials that enable satisfactory results to be achieved should be used.
(13)  There should be adequate service provisions for the fermenter.

The maintenance of an aseptic environment and the provision of aerobic conditions are probably the two most important criteria to be considered. The most widely used industrial scale fermenters are stirred, baffled, aerated tanks provided with systems of temperature, pH and foam formation control. A schematic representation of such a fermenter is shown in Fig. 8.1. Although the vast majority of fermenters in commercial operation rely on mechanical agitation to achieve good mixing and oxygen transfer some fermenters rely upon very high gas inputs without mechanical agitation. Fig. 8.2 illustrates some examples of non-mechanically stirred vessels, the most well-known of which is probably that used for the ICI Pruteen process.

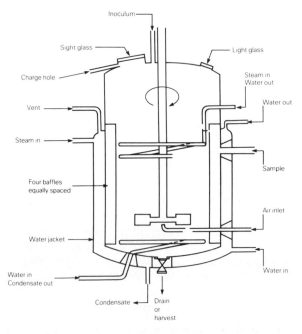

**Fig. 8.1**  Schematic representation of a stirred, aerated fermenter

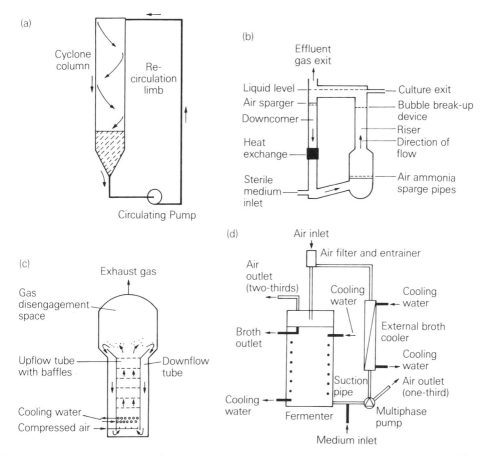

**Fig. 8.2** Some examples of non-mechanically agitated fermentation systems. (a) Cyclone fermenter. (Reproduced with permission from Dawson, 1974.) (b) Air-lift fermenter with external loop. (Reproduced with permission from Taylor and Senior, 1978.) (c) ICI air-lift fermenter with internal loop. (Reproduced with permission from Smith, 1980.) (d) Vogelbusch deep-jet fermenter. (Reproduced with permission from Hamer, 1979)

### 8.2.1 *Oxygen transfer*

Oxygen is a sparingly soluble gas (a saturated oxygen solution contains approximately $10 \text{ mg dm}^{-3}$ oxygen) and provision is achieved by sparging air into the vessel throughout the fermentation. The transfer of oxygen into solution is described by the equation:

$$\frac{dC_L}{dt} = K_L a (C^* - C_L) \qquad (1)$$

where $C_L$ is the dissolved oxygen concentration $(\text{mmol dm}^{-3})$
$\quad t$ is time (h)
$\quad C^*$ is the saturated dissolved oxygen concentration $(\text{mmol dm}^{-3})$
$\quad K_L$ is the mass transfer coefficient $(\text{cm h}^{-1})$

$a$ is the gas/liquid interface area per unit volume

$\dfrac{dC_L}{dt}$ is the oxygen transfer rate (mmol dm$^{-3}$ h$^{-1}$).

The terms $K_L$ and $a$ are very difficult to measure in a fermentation and, therefore, the two terms are combined in the expression $K_La$ or the volumetric transfer coefficient which is used as a measure of the aeration capacity of a fermenter. The value of $K_La$ is affected by the design of the vessel, the degree of agitation in a stirred vessel (measured by the amount of power consumed in stirring the vessel contents), the air flow rate, the viscosity of the culture broth and the presence of antifoam agents in the medium (Stanbury and Whitaker, 1984). Thus, the design of the fermenter and the operating conditions employed must be such that the organism's oxygen demand is met. However, it should be appreciated that the dissolved oxygen concentration in the medium has a marked influence on the physiology of an aerobic organism. Thus, the $K_La$ of the fermenter must be such that the desired amount of oxygen can be transferred per unit time (and therefore meet the oxygen demand of the organism) in the presence of a certain dissolved oxygen concentration in the medium (which maintains the cells in the correct physiological form).

The effect of dissolved oxygen concentration on the specific oxygen uptake rate (mmol of oxygen per g dry weight of biomass/h) is of the Michaelis-Menten (or Monod) type of relationship, as shown in Fig. 8.3. From Fig. 8.3 it may be seen that as the dissolved oxygen concentration is increased the specific oxygen uptake rate increases up to a maximum rate. The lowest dissolved oxygen concentration at which the specific oxygen uptake rate is maximum is termed $C_{crit}$. Thus, to maintain the culture in a fully aerobic state the fermenter must be capable of maintaining the dissolved oxygen concentration above $C_{crit}$, some representative values of which are given in Table 8.1. However, the optimum aeration conditions for microbial product formation may be different from those giving maximum biomass production. Thus, it may be necessary to operate a fermenter at a dissolved oxygen concentration either below or above $C_{crit}$.

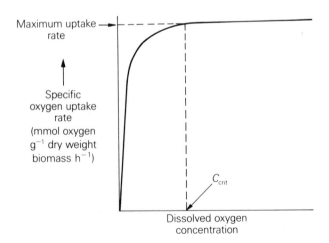

**Fig. 8.3**   The effect of dissolved oxygen on the specific oxygen uptake rate of a microbial culture

**Table 8.1** Critical dissolved oxygen concentrations of some representative micro-organisms (Rivere, 1977)

| Organism | Temperature | $C_{crit}$ (mmol dm$^{-3}$) |
|---|---|---|
| *Azotobacter* sp. | 30 | 0.018 |
| *Escherichia coli* | 37 | 0.008 |
| *Saccharomyces* sp. | 30 | 0.004 |
| *Penicillium chrysogenum* | 24 | 0.022 |

The production of commercial alcohol from yeast is an obvious example of the dissolved oxygen concentration being kept below $C_{crit}$. Although the production of alcohol is considered to be an anaerobic process, oxygen is still required for the synthesis of the sterol and unsaturated fatty acid components of the yeast membrane. Thus, the fermenter for such a system need only be designed for the provision of relatively low amounts of oxygen.

An excellent example of the effects of dissolved oxygen concentration on the production of amino acids is provided by the work of Hirose and Shibai (1980). These workers demonstrated that the $C_{crit}$ for *Brevibacterium flavum* was 0.01 mg dm$^{-3}$ and represented the degree of oxygen supply as the extent of oxygen 'satisfaction'—the respiratory rate of the culture expressed as a fraction of the maximum respiratory rate. Therefore, a value of oxygen satisfaction below 1 indicated that the dissolved oxygen concentration was below $C_{crit}$. The production of members of the glutamate and aspartate families of amino acids was affected detrimentally by oxygen satisfaction values of below 1, whereas optimum phenylalanine, valine and leucine productivities occurred at satisfaction levels of 0.55, 0.60 and 0.85 respectively. The biosynthetic routes of these amino acids are shown in Fig. 8.4, from which it may be seen that

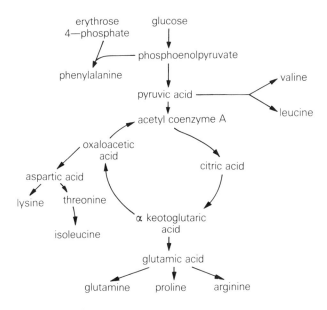

**Fig. 8.4** The biosynthetic routes to selected amino acids in *Brevibacterium flavum*

phenylalanine, valine and leucine are derived from the glycolysis intermediates, phosphoenolpyruvate and pyruvate, whereas the glutamate and aspartate families are derived from TCA cycle intermediates. Thus, oxygen limitation and the concomitant reduction in the operation of the TCA cycle would result in more intermediates being available for phenylalanine, leucine and valine biosynthesis. On the other hand, complete oxygen satisfaction results in a fully operational TCA cycle and, therefore, a ready supply of glutamate and aspartate family precursors.

The production of cephalosporin by *Cephalosporium* sp. (Feren and Squires, 1969) was optimum at dissolved oxygen concentrations considerably higher than the $C_{crit}$ value for the producing strain. Thus, in this case the fermenter would have to be capable of providing a higher dissolved oxygen concentration than that necessary to support the culture in aerobic conditions.

The rheological properties of the fermentation broth have a marked influence on the $K_L a$ achievable in the fermenter. A unicellular organism will normally give rise to a broth with a relatively low viscosity, unaffected by the degree of agitation in the vessel—that is, a Newtonian fluid. However, a mycelial organism will often produce a highly viscous broth, the viscosity of which varies with the degree of agitation—that is, a non-Newtonian fluid. The most common type of non-Newtonian mycelial broth is described as a pseudoplastic—that is, the viscosity of the broth decreases with increasing agitation. The degree of agitation in a stirred fermenter will vary through the volume of the vessel—the highest will occur in the region of the agitator, and the lowest in the regions most distant from the agitator. Thus, broth in the central zone of the fermenter will tend to be the least viscous which may result in the sparged air rising through this region of least resistance and anaerobic conditions developing in the more viscous regions. Therefore, it is important that in such a fermentation the agitation regime is one that results in good oxygen transfer in a region of high shear but also circulates the broth through the high shear zone. One approach to this problem is to use two different types of agitation in the system—one designed to produce a high shear zone, and the other to pump the broth. Legrys and Solomon (1977) designed a fermenter with a bottom-mounted disc turbine and a top-mounted curve blade impeller. The turbine produces a high degree of turbulence while the impeller produces a high flow velocity resulting in the circulation of one tank volume every 20–30 s. Thus, the mycelium is circulated through the aeration zone of the vessel before it becomes oxygen limited. Marsh and Pinkney (1985) described the pilot process for the cultivation of *Fusarium graminearum* used for the production of ICI-RHM mycoprotein. As already discussed (see Chapter 7.1) the filamentous structure of the organism is critically important to the eating quality of many of the products made from mycoprotein and, therefore, the fermenter must be capable of aerating a highly viscous non-Newtonian broth. The 1.3 m$^3$ pilot fermenter uses an impeller with a six-blade Rushton turbine producing oxygen transfer and a three-blade sabre giving rapid circulation within the vessel.

### 8.2.2 *Maintenance of aseptic conditions*
It is essential that a fermenter and its contents are sterilized prior to the fermentation and maintained aseptic throughout the fermentation period. The achievement of aseptic conditions is, fundamentally, an engineering problem and is therefore outside the scope of this text, but the major points will be summarized to familiarize the reader with this important area.

   (1) In an agitated vessel the agitator drive shaft must enter the vessel via an aseptic seal.

(2) The fermenter must be constructed such that it, and its contents, may be steam-sterilizable. Thus steam must be supplied at each entry or exit point of the vessel. During sterilization steam is introduced through all the entry and exit points, with the exception of the air outlet from which steam is allowed to escape.

(3) The air supply should be filter sterilized.

(4) During the fermentation the vessel should be operated at a positive pressure so that any leakage results in air leaving the fermenter rather than contaminated air from outside being drawn in.

(5) The vessel should be designed such that it may be inocculated, sampled and additions made aseptically.

The operations of sterilization and aseptic addition on an industrial plant are complex procedures involving the manipulation of, perhaps, hundreds of valves. Thus, the probability of human error is relatively large. However, the advent of computer control has proved to be an enormous asset in the control and monitoring of sterilization and aseptic operations. According to Tonge (1980) the operation of the ICI Pruteen plant would be extremely difficult without computer control.

## 8.3 The design of the culture medium

The culture medium which is used in a fermentation must not only meet the requirements of the micro-organism, but also those of the industrial process. Thus, such factors as its cost, efficiency of utilization, rheology and its effect on downstream processing are all important factors in the design of a fermentation medium.

All micro-organisms require water, sources of carbon, nitrogen, mineral elements and, possibly, specific requirements such as vitamins and amino acids. The vast majority of commercially significant micro-organisms are chemo-organotrophic and, therefore, the energy and carbon sources are one and the same. Carbon and nitrogen are usually supplied as complex mixtures of cheap natural products or byproducts of some process. The trace elements are often present in the complex components as well as in the tap water used to formulate the medium. Any specific requirements, such as amino acids or vitamins, may be added in pure form but it is more likely that will be supplied more economically as a complex plant or animal extract. Phosphates may be incorporated into the medium as buffering agents although external pH control is normally used. Table 8.2 includes some examples of carbon and nitrogen sources commonly used in fermentation media and Table 8.3 includes some examples of industrial media formulations.

### 8.3.1 *Yield of biomass*
The yield of biomass obtainable per unit of carbon incorporated into the medium is an important criterion in the choice of carbon source, especially for single-cell protein processes. The cellular yield coefficient ($Y$) for a carbon source is described as:

$$\frac{\text{The quantity of dry cell matter produced}}{\text{The quantity of carbon substrate utilized}}$$

Some representative values of $Y$ are given in Table 8.4 from which it may be seen that some hydrocarbons give better yield coefficients than carbohydrates. The reason for

**Table 8.2**   Some examples of carbon and nitrogen sources commonly used in fermentation media

| *Carbon sources* | *Nitrogen sources* |
|---|---|
| Starch | Ammonia |
| Glucose | Ammonium salts |
| Malt | Nitrates |
| Beet molasses | Corn-steep liquor |
| Cane molasses | Peanut granules |
| Vegetable oils | Soybean meal |
| Hydrocarbons | Dried blood |
| | Pharmamedia |
| | Soya meal |
| | Distillers' solubles, |
| | Yeast extract |

**Table 8.3**   Some examples of industrial fermentation media

| *Glutamic acid production* (Gore, Reisman and Gardner, 1968) | | *Clavulanic acid* (Butterworth, 1984) |
|---|---|---|
| Dextrose | $270 \text{ g dm}^{-3}$ | Soybean flour 1.5% |
| $(NH_4)_2HPO_4$ | $2 \text{ g dm}^{-3}$ | Glycerol 1.0% |
| $NH_4H_2PO_4$ | $2 \text{ g dm}^{-3}$ | $KH_2PO_4$ 0.1% |
| $K_2SO_4$ | $2 \text{ g dm}^{-3}$ | 10% Pluronic L81 |
| $MgSO_4 \cdot 7H_2O$ | $0.5 \text{ g dm}^{-3}$ | Antifoam in |
| $MnSO_4 \cdot H_2O$ | $0.04 \text{ g dm}^{-3}$ | soybean oil 0.2% |
| $FeSO_4 \cdot 7H_2O$ | $0.02 \text{ g dm}^{-3}$ | (vol/vol). |
| Biotin | $12 \text{ }\mu\text{g dm}^{-3}$ | |
| Penicillin | $11 \text{ }\mu\text{g dm}^{-3}$ | |

| *Griseofulvin* (Huber and Tietz, 1984) | |
|---|---|
| Corn-steep liquor | 3.50 g/l |
| $(NH_4)_2SO_4$ | 0.50 g/l |
| $KH_2PO_4$ | 4.00 g/l |
| KCl | 1.00 g/l |
| $CaCO_3$ | 4.00 g/l |
| $H_2SO_4$ | 0.125 g/l |
| Mobilpar | 0.275 g/l |
| White mineral oil | 0.275 g/l |

Glucose (50%) fed at $0.5–0.6 \text{ dm}^3 \text{ h}^{-1}$ to maintain pH 6.8–7.2

**Table 8.4** Some examples of yield factors for a range of carbon sources and micro-organisms. (Modified from Mateles, 1979)

| Substrate | Organism | $Y$ (g biomass $g^{-1}$ substrate) | $O_2$ requirement (g$O_2$ per g dry wt.) | Reference |
|-----------|----------|-----------------------------------|------------------------------------------|-----------|
| Glucose | *Escherichia coli* | 0.53 | 0.4 | Schulze and Lipe (1964) |
| Methanol | *Pseudomonas* c. | 0.54 | 1.2 | Goldberg *et al* (1976) |
| Octadecane | *Pseudomonas* sp. | 1.06 | 1.6 | Wodzinski and Johnson (1968) |

this is that the highly reduced hydrocarbons contain more carbon per g of compound than do the carbohydrates. However, the improved yield is 'paid for' by the high oxygen demand of cells growing on these highly reduced carbon compounds, also shown in Table 8.4. Therefore, far more oxygen has to be provided to a hydro-carbon-based fermentation than to a carbohydrate-based one. Also, the consumption of large amounts of oxygen results in the generation of considerable heat which must be removed to maintain the temperature of the fermentation at the optimum value.

Thus, the choice of carbon source has a major effect on the design of the fermentation vessel which has to provide sufficient oxygen for complete utilization of the carbon source and a sufficient heat exchange area to remove the heat produced. An excellent example of this is the ICI Pruteen fermenter (Smith, 1980).

### 8.3.2 *Inducers*

Besides providing substrate for the growth of the process organism the medium will have a very considerable influence on the products produced by the organism. Catabolic enzymes are frequently inducible, that is they are only produced in the presence of an inducer which is normally the substrate of the enzyme or a structurally similar compound. Thus, a medium for the commercial production of a catabolic enzyme must include the relevant inducer. Table 8.5 cites some examples of commercially relevant enzymes and inducer systems.

The production of certain secondary metabolites has also been shown to be subject to control by induction. For example, $\alpha$-mannosidase is a key enzyme in the

**Table 8.5** Some examples of inducers used in the production of commercially relevant enzymes

| Enzyme | Inducer |
|--------|---------|
| Proteases | Various proteins |
| $\alpha$-Amylase | Starch |
| Cellulase | Cellulose |
| Pectinase | Pectin |
| Penicillin acylase | Penylacetic acid |

biosynthesis of streptomycin and is induced in the presence of yeast mannan. Thus, yeast extract is often used as a nitrogen source in the streptomycin fermentation.

### 8.3.3  *Repressors*

The production of catabolic enzymes and secondary metabolites is frequently repressed by the presence of certain compounds in the culture medium. Rapidly utilized carbon sources have been demonstrated to repress the formation of amylases and a wide range of secondary metabolites, for example, griseofulvin, penicillin, bacitracin and actinomycin (Demain, 1984). After many years of empirical medium development most antibiotics are produced using carbon sources other than D-glucose, or, if D-glucose is used it is fed to the culture at a low rate (as discussed in Section 8.4).

Nitrogen sources have also been demonstrated to have repressive effects on secondary metabolite production. For example, the fact that soybean meal and proline are the best nitrogen sources for streptomycin production is probably due to their slow utilization, thus avoiding nitrogen metabolite repression (Demain, 1984). Another common repressor of secondary metabolism is inorganic phosphate as demonstrated in the streptomycin, bacitracin, oxytetracycline and novobiocin fermentations.

Thus, the choice of the limiting factor (the component which is exhausted first and therefore limits growth) in a fermentation medium will be determined by the repressive effects of the medium constituents on productivity. Therefore, depending on the control system involved, commercial media may be carbon, phosphate or nitrogen limited or, occasionally, limited by some other repressing nutrient component.

### 8.3.4  *Precursors*

The productivity of a certain metabolite may be increased considerably if a precursor of the metabolite is fed to the system. The classical example of this situation is the addition of phenylacetic acid to the penicillin fermentation which acts as the precursor of the side chain of benzylpenicillin. Further examples are provided by the precursor role of chloride ion in the chlortetracycline and griseofulvin fermentations.

## 8.4  The mode of operation of fermentation processes

As discussed earlier in Chapter 7 micro-organisms may be grown in batch, fed-batch and continuous culture. Although continuous culture offers the greatest degree of control over the growth and physiology of the cells, its use by the fermentation industry is extremely limited, for the reasons outlined previously (see Section 7.4). However, the adoption of fed-batch culture has given the fermentation technologist a valuable tool for the control of the environment of the fermentation. The most common type of fed-batch system employed in mass culture is to feed one component of the medium to the fermentation, the feed rate being controlled by some measurable parameter of the fermentation. The most common substrate to be fed is the carbon source but any component that has a critical controlling effect on the fermentation may be used. The most common measurable parameters used to control the feed rate are dissolved oxygen concentration and pH, although off-line analyses such as viscosity may be used as control parameters. The major advantage of feeding a medium component to a culture, rather than incorporating it entirely in the initial batch, is that the nutrient may be maintained at a very low concentration during the fermentation. A low (but constantly replenished) nutrient level may be advantageous in:

(1) Maintaining conditions in the culture within the aeration capacity of the fermenter.
(2) Removing the repressive effects of medium components such as rapidly used carbon, nitrogen and phosphate sources.
(3) Avoiding the toxic effects of an essential medium component.
(4) Providing a limiting level of a required nutrient for an auxotrophic mutant.

The earliest example of the commercial use of fed-batch culture is the production of bakers' yeast. It was recognized as early as 1915 that an excess of malt in the production medium would result in a high rate of biomass production and an oxygen demand which could not be met by the fermenter (Reed and Peppler, 1973). This resulted in the development of anaerobic conditions and the formation of ethanol at the expense of biomass. The solution to this problem was to grow the yeast initially on a weak medium and then add additional concentrated medium at a rate less than the organism could use it. It is now appreciated that a high glucose concentration represses respiratory activity and in modern yeast production plants the feed of molasses is under strict control based on the automatic measurement of traces of ethanol in the exhaust gas of the fermenter. As soon as ethanol is detected the feed rate is reduced. Although such systems result in lower growth rates than the maximum obtainable the biomass yield is nearly the theoretical obtainable (Fiechter, 1982).

The penicillin fermentation provides a very good example of the use of fed-batch culture for the production of a secondary metabolite (Hersbach, Van der Beek and Van Dijck, 1984). The penicillin fermentation is a 'two-stage' process; an initial growth phase is followed by the production phase or idiophase. During the production phase glucose is fed to the fermentation at a rate which allows a relatively high growth rate (and therefore rapid accumulation of biomass), yet limits the oxygen demand to that which can be met by the fermenter. The feed rate is controlled by either the dissolved oxygen concentration or the pH of the broth. If the dissolved oxygen concentration drops below $C_{crit}$ (see Section 8.2) the flow rate is decreased. If pH is used to monitor the system a decline in the pH value would indicate the production of organic acids and, therefore, anaerobic conditions. Thus, a decline in the pH dictates a decrease in the glucose addition rate. During the production phase the biomass must be maintained at a relatively low growth rate and, thus, the glucose is fed at a low rate which keeps it below the concentration which represses penicillin formation (see Section 8.3.3). During this phase the dissolved oxygen concentration is used as a 'feedback' control of the addition rate. Phenylacetic acid is a precursor of penicillin G (benzylpenicillin) (see Section 8.3.4) but is also toxic to the fungus above a threshold concentration. Thus, phenylacetic acid is also fed to the fermentation during the production phase thereby maintaining its concentration below the toxic level.

In Chapter 9 the use of auxotrophic mutants is discussed for the production of microbial products. Such mutants are capable of growth only when supplied with certain compounds not needed by the wild-type cells. Furthermore, to achieve optimum productivity of the desired product the growth of an auxotrophic mutant should be limited by the availability of the required compound. Thus, the feeding of the required compound to the fermentation at a rate less than that at which it may be consumed enables the fermentation technologist to control the process within relatively narrow margins.

## 8.5   Summary

The performance of a process organism is affected markedly by its environment. The environment of the organism is influenced by the mode of operation, and design of both the fermenter and the culture medium. A fermenter is designed to maintain the culture in a pure state under optimum physical and chemical conditions throughout the fermentation. The provision of oxygen to the organism must be such that product formation is optimum which may mean that the dissolved oxygen concentration has to be different from that supporting optimum growth. The medium used in the fermentation should contain all necessary requirements for growth as well as additives that may be needed to support optimum product synthesis. The environment of the organism may be influenced considerably by the mode of operation of the fermentation, the most common type of which is fed-batch culture. Fed-batch systems allow the fermentation technologist to maintain aerobic conditions, to control the repressing effects of medium components and to provide essential, but sometimes toxic materials, in low concentrations. Thus, the maintenance of a controlled environment requires a combination of biological and chemical engineering expertise.

# Chapter 9

# *The improvement of industrial micro-organisms*

## 9.1 Introduction

The ideal industrial micro-organism is one that produces the desired commercial metabolite in vast quantities relative to its other biosynthetic and catabolic end products. Obviously, such an industrial strain would be a hopeless competitor in the natural environment. Thus, organisms isolated from the natural environment normally produce a desired metabolite in very small amounts and, therefore, it is necessary to improve the productivity of the isolate. Although product yield may be improved by optimizing the cultural conditions, productivity is ultimately controlled by the genome. Thus, to improve the potential productivity, the organism's genome must be modified and this may be achieved in two ways: (a) mutation, or (b) recombination.

## 9.2 Mutation

Each time a microbial cell divides there is a small probability of an inheritable change occurring. A strain exhibiting such a changed characteristic is termed a mutant, and the process giving rise to it a mutation. The probability of a mutation occurring may be increased by exposing the culture to a mutagenic agent such as ionizing radiation, ultraviolet light and various chemicals, for example, nitrous acid and nitrosoguanidine. Such an exposure usually involves subjecting the cells to a mutagenic dose which results in the death of the vast majority of the cells. The survivors of this exposure may then contain some mutants, a very small proportion of which may be superior producers. It is not possible, using standard mutation techniques, to predetermine the gene that will be affected by the mutagen because the particular sites of action of mutagens will occur in a very large number of genes. Therefore, it is the task of the industrial geneticist to differentiate the few superior producers (the desirable types) from the many inferior producers found among the survivors of a mutation treatment.

The task is far easier for strains producing primary metabolites than it is for those producing secondary metabolites, as may be seen from the following examples.

### 9.2.1   *The selection of mutants producing improved levels of primary metabolites*

One of the most important groups of commercial primary metabolites is the amino acids, and the most common organism for their production is *Corynebacterium glutamicum*. The original *C. glutamicum* was isolated by Kinoshita, Udaka and Shimono (1957) who screened natural isolates for the ability to excrete glutamic acid into the medium. It is relevant at this stage to consider the physiology of glutamic acid production by *C. glutamicum*. Kinoshita's natural isolate was a biotin auxotroph—that is, it was unable to synthesize biotin which then had to be added to the growth medium. When *C. glutamicum* was grown in a medium containing a high concentration of biotin the organism synthesized glutamate at a level of 25–36 µg mg$^{-1}$ dry weight of cells. Further production is assumed to have been prevented by some form of feedback control by glutamate on its own synthesis. Under biotin limiting conditions (that is, the growth of the cells was limited by the availability of biotin) glutamate was excreted from the cells and accumulated in the medium up to a concentration of 50 g dm$^{-3}$. The explanation of the excretion of glutamate is based upon the fact that biotin limitation results in the cell membranes being deficient in phospholipids thus disrupting their selective permeability (Nakao *et al.*, 1973).

However, Kinoshita's isolate was not only deficient in the ability to synthesize biotin but also in the ability to synthesize the enzyme $\alpha$-ketoglutarate dehydrogenase which converts ketoglutarate to succinate in the tricarboxylic acid cycle. This deficiency in the TCA cycle results in $\alpha$-ketoglutarate being diverted to glutamate synthesis. The deficiency in the TCA cycle is made good by the operation of the glyoxylate pathway in this organism, as shown in Fig. 9.1.

Thus, the combination of a metabolic block in the conversion of a TCA cycle intermediate and biotin limitation results in very high glutamate excretion. A major disadvantage of the use of biotin limitation as a control factor is that this precludes the use of biotin-rich, crude carbohydrate carbon sources. An alternative approach that has been used is to include some factor in the growth medium which disrupts the permeability of the cells, and in such cases the biotin concentration need not be rigidly controlled. Examples of such 'disruptive factors' are penicillin and surfactants such as fatty acid derivatives.

Mutants of *C. glutamicum* have been used extensively for the commercial production of a number of amino acids (as well as glutamic acid). The synthesis of primary biosynthetic end products is controlled in wild-type strains such that they are produced only at a level required by the organism. The control mechanisms involved are the inhibition of the activity of key enzymes in the biosynthetic route and the repression of their synthesis by the end product(s) when present in the cell at a concentration sufficient to meet the organism's requirements. Thus, these mechanisms are referred to as feedback control. *C. glutamicum* is well suited to industrial utilization because feedback control is relatively simple and, therefore, may be eliminated with greater ease than that of more 'sophisticated' organisms such as *E. coli*, for example. The isolation of mutants of *C. glutamicum* (and closely related organisms) for the production of lysine will be used as examples to illustrate the approaches that have been adopted to remove biosynthetic control.

The control of lysine production in *C. glutamicum* is illustrated in Fig. 9.2, from which it may be seen that aspartokinase, the first enzyme in the biosynthetic route, is

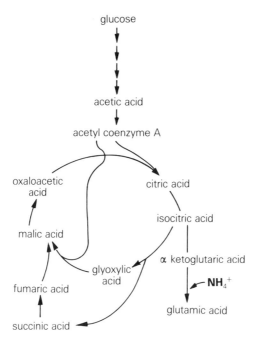

**Fig. 9.1**  The TCA cycle and glyoxylate cycle in *Corynebacterium glutamicum*

inhibited only when both lysine and threonine are present above a threshold level. This type of control is termed concerted feedback inhibition. An important feature of the pathway is that lysine does not exert any control over the biosynthetic route from aspartic semialdehyde to lysine. A mutant which could not catalyse the conversion of

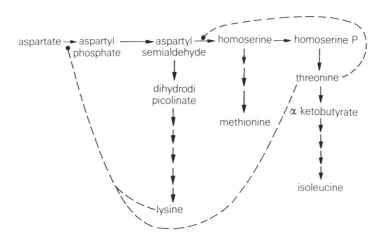

**Fig. 9.2**  The control of lysine production in *Corynebacterium glutamicum*;  – – – – – feedback inhibition

aspartic semialdehyde to homoserine would be capable of growth only in homo-serine-supplemented medium and the organism would be described as a homoserine auxotroph. If such an organism were grown in the presence of very low concentrations of homoserine the endogenous level of threonine would not reach the inhibitory concentration for aspartokinase control, and aspartate would be converted to lysine which would accumulate in the medium. Thus, a knowledge of the control of the biosynthetic pathway allows a 'blueprint' of the desirable mutant to be constructed and makes the task of designing the procedure to separate the desired type from the other survivors of a mutation treatment much easier. Using this logical approach Nakayama, Kituda and Kinoshita (1961) isolated homoserine auxotrophs of *C. glutamicum* using the penicillin enrichment technique developed by Davis (1949). Under normal cultural conditions an auxotroph is at a disadvantage compared with the parental (wild-type) cells. However, pencillin only kills growing bacterial cells and, therefore, if the survivors of a mutation treatment were cultured in a medium containing penicillin and lacking the growth requirement of the desired mutant only those cells unable to grow would survive, that is, the desired auxotrophs. If the cells were removed from the pencillin broth, washed and resuspended in a medium containing the requirement of the desired auxotroph then the resulting culture should be rich in the required type. Nakayama's group succeeded in isolating a homoserine auxotroph of *C. glutamicum* which synthesized $44 \, \text{g} \, \text{dm}^{-3}$ lysine.

An alternative approach to the isolation of mutants which do not produce controlling end products (auxotrophs) is to isolate mutants which do not recognize the presence of the controlling factors. Such mutants may be isolated from the survivors of a mutation treatment by exploiting their capacity to grow in the presence of certain compounds which are inhibitory to the parental types. An analogue is a compound similar in structure to another compound, and analogues of primary metabolites are frequently inhibitory to microbial cells. The toxicity of an analogue may be due to any of a number of possible mechanisms; for example, the analogue may be incorporated into a macromolecule in place of the natural product resulting in the production of a defective compound; or the analogue may act as a competitive inhibitor of an enzyme for which the natural product is a substrate. Also, the analogue may mimic the control characteristics of the natural product and inhibit the formation of the product despite the fact that the natural product concentration is inadequate to support growth. A mutant which is capable of growth in the presence of an analogue inhibitory to the parent may owe its resistance to any of a number of mechanisms. However, if the toxicity were due to the analogue mimicking the control characteristics of the normal end product then the resistance may be due to the control system being unable to recognize the analogue as a control factor. Such analogue resistant mutants may also not recognize the natural product and may, therefore, overproduce it. Thus, there is a reasonable probability that mutants resistant to the inhibitory effects of an analogue may overproduce the compound to which the analogue is analogous.

Sano and Shiio (1970) made use of this approach in attempting to isolate mutants of *Brevibacterium flavum*. The control of lysine production in *B. flavum* is the same as that shown for *C. glutamicum* in Fig. 9.2. Sano and Shiio demonstrated that the lysine analogue s-(2-aminoethyl) cysteine (AEC) only inhibited growth completely in the presence of threonine which indicates that AEC, combined with threonine, caused concerted feedback inhibition of aspartokinase and deprived the organism of lysine and methionine. Mutants capable of growing in the presence of AEC and threonine were isolated by plating the survivors of mutation treatments onto agar containing the two factors. A relatively high proportion of the resulting colonies were lysine overproducers, the best of which produced more than $30 \, \mathrm{g \, dm^{-3}}$.

Organisms used for the commercial production of primary metabolites are rarely modified at only one genetic site. It is frequently necessary to alter several control sites to achieve overproduction of the desired compound; for example Kase and Nakayama (1972) produced a strain of *C. glutamicum* that overproduced threonine which was auxotrophic for methionine and resistant to the threonine analogue $\alpha$-amino-$\beta$-hydroxy valeric acid and to the lysine analogue s-($\beta$-aminoethyl)-L-cysteine. However, multiply-mutated strains are also used for the commercial production of compounds where a mutation at only one site should give good productivity. The objective of using such strains is to improve their stability in that more than one mutation giving the *same* phenotype means that there must be more than one reversion event before the cell looses its commercially important characteristic. For example, Sano and Shiio (1970) developed a *C. glutamicum* strain for the production of lysine which was auxotrophic for homoserine and leucine and was resistant to the analogue s-(2-aminoethyl)-L-cysteine.

### 9.2.2 *The selection of secondary metabolite producing mutants*

A knowledge of the control systems, therefore, may assist in the design of procedures for the isolation of mutants overproducing primary metabolites. The design of procedures for the isolation of mutants overproducing secondary metabolites is more difficult due to the fact that far less information is available on the control of production and, also, that the end products of secondary metabolism are not required for growth. The systems that have evolved, and achieved considerable success, are direct empirical screens of the survivors of a mutation treatment for productivity rather than cultural systems which give an advantage to potentially producing types. Thus, a typical screen for improved productivity mutants would be to subject a population of cells to a mutation treatment such that 1–5% of the cells survived and to screen as many of the survivors as possible for productivity. The assessment of productivity is usually performed by culturing the survivors in shake flasks and assaying the culture for activity. These procedures are obviously very labour-intensive, and various attempts have been made to miniaturize the systems such that the survivors are cultured and assayed for productivity on agar media. Such miniaturized systems should be far less labour-intensive than the conventional programmes; examples of these systems are given in Table 9.1.

Although the empirical screens for improved secondary metabolite producers have met with considerable success there is an increasing tendency to utilize some form of selective culture in the isolation of desirable mutants. Such systems include the isolation of auxotrophs, analogue-resistant mutants and those resistant to the autotoxic effects of the secondary metabolite. Several workers have obtained improved secondary metabolite producing strains by isolating auxotrophic mutants. In many of these cases there appears to be no correlation between the compound and the secondary product synthesized. One possible explanation of this effect is that the improved producers were double mutants and that their auxotrophy was not directly related to their improved productivity. However, such systems may be exploited using nitrosoguanidine (NTG) as a mutagen. NTG causes clusters of mutations around the replicating fork of the bacterial chromosome. Thus, if one of the mutations were selectable (for example, by auxotrophy) it may be possible to isolate a strain containing the selectable mutation along with the non-selectable ones which map close

**Table 9.1** Miniaturized systems developed for the screening of improved secondary metabolite producers

| Secondary product | Reference |
|---|---|
| Chlortetracycline | Dulaney and Dulaney (1967) |
| Kasugamycin | Ichikawa *et al*. (1971) |
| Penicillin | Ditchburn, Giddings and MacDonald (1974) |
| Penicillin | Ball and MacGonagle (1978) |
| Cephalosporin | Trilli *et al*. (1978) |
| β-Lactam Antibiotics | Ray Chowdhurry, Goswani and Chakrabarti (1980) |

by. The efficient use of this approach would require an accurate knowledge of the positions of genes important in secondary metabolism so that neighbouring mutations may be selected.

The technique of selecting mutants resistant to inhibitory analogues has found some application in the selection of secondary metabolite overproducers. If the availability of a secondary metabolite precursor is a limiting factor to productivity of the secondary metabolite, then it may be possible to stimulate production by improving synthesis of the precursor which is frequently an end product of primary metabolism. For example, Elander *et al*. (1971) isolated tryptophan analogue-resistant mutants of *Pseudomonas aureofaciens* which overproduced the antibiotic pyrrolnitrin. Tryptophan is a precursor of pyrrolnitron and the resistant mutant could synthesize more of this limiting precursor.

Martin *et al*. (1979) isolated tryptophan analogue-resistant mutants which overproduced candicidin. The explanation of this phenomenon is that tryptophan feedback inhibits its own synthesis which results in the depletion of chorismic acid—a precursor of both tryptophan and candicidin. In the resistant mutant tryptophan did not feedback inhibit its formation and therefore the production of the candicidin precursor was not depleted.

Many secondary metabolites have been demonstrated to inhibit the producer strain if they are added during the growth phase of the organism. Also, there appears to be a correlation between the autotoxic resistance in the growth phase and productivity in the idiophase. Thus, the ability of a survivor of a mutation treatment to grow in the presence of a high concentration of its secondary metabolite may be used as a selectable characteristic for improved productivity. The approach has been used successfully by Bu'Lock (1980) in the isolation of strains overproducing the antifungal antibiotics, agestorols. The technique has been taken further using genetic manipulation methods by Crameri, Davies and Thompson (1985). These workers cloned the gene coding for 6′N-actyltransferase into *Streptomyces kanamyceticus*, a kanamycin producer. The enzyme is involved in aminoglycoside resistance and the recombinants containing the cloned gene were more resistant to, and produced more of, kanamycin. Thus, the application of the techniques used in the improvement of primary metabolite formation has resulted in considerable progress being made in the improvement of secondary metabolite producers.

## 9.3 Recombination

Hopwood (1979) defined recombination as any process which helps to generate new combinations of genes that were originally present in different individuals. Compared with the use of mutant induction and selection the use of recombination for industrial strain improvement has been fairly limited. This is probably due to the success of the mutation programmes, their relative ease and the lack of basic genetic information on industrial strains. However, techniques are now widely available which allow the use of recombination as a system for industrial strain improvement. Relatively few industrially important organisms exhibit sexual reproduction as such, and the attainment of recombinants of such types is probably confined to commercial mushrooms and the yeast strains used in the baking, brewing and distilling industries. However, recombination systems not associated with sexual reproduction are more common in industrially important strains, and recently developed experimental methodologies have made these systems easier to exploit. Some of these recombination systems will be described briefly followed by an account of the techniques of protoplast fusion which have greatly increased the probability of obtaining worthwhile recombinants.

### 9.3.1 *Recombination systems applicable to commercially relevant organisms*

Pontecorvo and co-workers (1953) demonstrated that nuclear fusion and gene segregation could take place outside the sexual organs of the sexual fungus, *Aspergillus nidulans*, and in the imperfect fungi (not displaying sexual reproduction), *Aspergillus niger* and *Penicillium chrysogenum*. The process was termed the parasexual cycle. For parasexual recombination to occur in an imperfect fungus nuclear fusion must occur between genetically unlike nuclei, that is, genetically unlike nuclei must be present in the one organism. Such an organism is termed a heterokaryon, and it may be formed by nuclei migrating from one individual to another via connections between the two individuals formed by the fusion of adjacent cells. The migrated nuclei may then divide in the recipient strain and coexist with the original nuclei of that strain. The establishment of a heterokaryon is a rare event and to aid its recognition auxotrophic markers are used. The two strains to be used to generate the heterokaryon are made auxotrophic for different requirements. Spores of both auxotrophs are mixed and spread onto agar medium lacking the specific requirements of each strain. Sufficient nutrients should be present in the spores to allow germination to take place but significant growth may only occur if vegetative cellular fusion takes place and a heterokaryon is produced. Thus, the colonies produced from such a procedure should be heterokaryons in which the nuclei of the two starting strains coexist.

Nuclear fusion may occur between unlike nuclei in the heterokaryon and give rise to a diploid clone. A diploid nucleus in the clone may, in rare cases, undergo an abnormal mitosis resulting in the production of a recombinant. The abnormal mitosis may involve mitotic crossing-over, haploidization or a combination of both. Mitotic crossing-over involves the exchange of distal segments between chromatids of homologous chromosomes and haploidization is a process which results in the unequal distribution of chromatids between the daughter nuclei of a mitotic division.

The frequency of vegetative nuclear fusion may be increased by the use of agents such as camphor vapour or ultraviolet light and the frequency of mitotic crossing-over, and haploidization may be increased using such agents as X-rays, ultraviolet light and nitrogen mustard. A full account of the parasexual cycle is given by Sermonti (1969). Thus, using these techniques recombinants may be obtained displaying characteristics originally found in the different parents. The technique has been used to study the basic genetics of *Penicillium chrysogenum* and *Aspergillus niger* as discussed by Sermonti (1969) and Macdonald and Holt (1976); Queener and Swartz (1979) have claimed that commercial penicillin producers have used the parasexual cycle to 'breed' improved commercial strains. One of the major difficulties inherent in the utilization of the parasexual cycle for strain improvement is the establishment of the heterokaryon, that is the fusion of cells from different strains, but the techniques of protoplast fusion have done much to alleviate this problem and this is discussed in a later section.

Conjugation is the process in bacteria whereby genetic information is transferred from one cell to another by cell-to-cell contact. The chromosome of the 'donor' cell is mobilized by the integration of a normally extrachromocomal DNA particle into the chromosome. In *E. coli* the 'mobilization' element is termed the F factor or plasmid whereas in other bacteria different factors are involved. The amount of genetic information which is transferred to the recipient cell is dependent on the time of contact between the two participating cells—thus, the longer the cells are in contact the greater amount of material will be transferred. Furthermore, the DNA is transferred in a linear fashion; a single strand of the donor DNA is cut adjacent to the site of integration of the mobilizing factor and this strand passes into the recipient cell, the mobilizing factor being the last element to be transferred. DNA synthesis is initiated in the donor strain such that the transferred DNA is replaced by newly synthesized material. The transferred segment may be incorporated into the recipient cell's chromosome by a crossing-over procedure. Thus, the transfer of a particular gene will depend on its position on the chromosome relative to the site of integration of the mobilizing factor and the period of time the cells are in contact. As in the case of the parasexual cycle the occurrence of conjugation is a rare event and, therefore, a selection procedure must be used to isolate the recombinants. This may be achieved by introducing different auxotrophic requirements into the strains and culturing the potential recombinants on minimal medium. However, because the transfer of genetic information is time-dependent the genetic markers used will result in the selection of only certain recombinants. Thus, it is important to use a range of auxotrophic markers which are situated on different regions of the chromosome and it is therefore essential to have a genetic map of the organism.

Conjugation has been demonstrated in the streptomycetes which have enormous industrial significance. Hopwood (1976) has claimed that industrial companies have used conjugation in strain improvement of industrial strains of streptomycetes but the results have not been published for obvious security reasons. However, the disadvantages of the conjugation system are that considerable genetic knowledge of the organism is required to perform the cross effectively and that, generally, incomplete transfer of the genome is obtained. Again, as in the case of the parasexual cycle the development of protoplast fusion techniques has reduced the enormity of these difficulties.

### 9.3.2    *Protoplast fusion*

Protoplasts are cells devoid of their cell walls and may be prepared by subjecting cells to the action of wall-degrading enzymes in isotonic solutions. Cell fusion, followed by nuclear fusion, may occur between protoplasts of cells that would not otherwise fuse and the resulting fused protoplast may regenerate a cell wall and grow as a normal cell. Protoplast fusion has been achieved with the filamentous fungi (Ferenczy, Kevei and Zsolt, 1974), yeasts (Sipiczki and Ferenczy, 1977), *Bacillus* sp. (Fodor and Alfoldi, 1976), *Brevibacterium flavum* (Tosaka *et al.*, 1982) and the streptomycetes (Chater, 1979).

Protoplast fusion has proved to be an enormous advantage in generating heterokaryons of the filamentous fungi—the most difficult task in the use of the parasexual cycle. The frequency of protoplast fusion in the filamentous fungi still necessitates the use of genetic markers to isolate the recombinants, but the degree of success in their isolation appears to have been far greater than the conventional techniques used for the generation of heterokaryons. Several examples of the use of the technique with commercially relevant strains exist in the literature. For example, Hamlyn and Ball (1979) fused strains of *Cephalosporium acremonium*; a cross between an asporulating, slow-growing strain, with a sporulating fast-growing strain which synthesized only one-third of the level of cephalosporin of the first strain; this resulted in the selection of a recombinant which sporulated well, had a high growth rate and synthesized the higher level of cephalosporin. Chang, Terasaka and Elander (1982) used protoplast fusion to combine the desirable properties of two strains of penicillin V-synthesizing strains of *Penicillium chrysogenum*. Strain 1 displayed an adverse morphology but synthesized only penicillin V (the desired product), whereas strain 2 displayed a desirable morphology yet synthesized significant quantities of penicillin OH-V (an undesirable product). The use of protoplast fusion resulted in the isolation of recombinants capable of synthesizing only penicillin V with the desired morphological type.

Protoplast fusion has also considerable advantages in the isolation of recombinants of the streptomycetes. Hopwood and co-workers (Hopwood, 1979) have developed a protoplast fusion technique for streptomycetes which gives a very high number of recombinants such that it may no longer be necessary to utilize auxotrophic markers to detect the recombinants. Furthermore, the whole of the genome is subject to recombination and not just a portion of it as is the case with conjugation. An excellent example of the use of the technique for the improvement of an amino acid producing bacterium is provided by the work of Tosaka *et al.* (1982). These workers improved the rate of glucose consumption (and therefore lysine production) of a high lysine-producing strain of *Brevibacterium flavum* by fusing it with another *B. flavum* strain which was a non-lysine producer but consumed glucose at a high rate.

The examples quoted of the use of protoplast fusion have all involved the combination of desirable properties, expressed in different strains, into one organism. Furthermore, the desirable properties were not confined to yield of the required product but included other characteristics of the organism that were important to the fermentation process—for example, the growth rate, degree of sporulation and rate of substrate consumption. Thus, recombination is an excellent tool to improve characteristics other than simply product yield.

### 9.3.3   In vitro *recombinant DNA technology*

*In vitro* recombination in industrially important micro-organisms has been achieved using the techniques of *in vitro* recombinant DNA technology discussed in Chapters 10 to 12. Although the most well-publicized recombinants achieved by these techniques are those bacterial and yeast strains which synthesize foreign compounds, very considerable achievements have been made in the improvement of strains producing conventional products. However, the vast majority of the work published is concerned with the improvement of primary-metabolite producing strains.

The efficiency of the organism used in the ICI single-cell protein process, *Methylophilus methylotrophus*, has been improved by the incorporation of a plasmid containing the glutamate dehydrogenase gene from *E. coli*. The manipulated strain was capable of more efficient ammonia metabolism than the original type which was reflected in a 5% improvement in carbon conversion. However, it appears that the manipulated organism has yet to be used on a commercial scale. The production of threonine by *E. coli* has been improved using *in vitro* manipulation techniques. Debabov (1982) incorporated the entire threonine operon of a threonine analogue-resistant mutant of *E. coli* $K_{12}$ into a plasmid which was then introduced back into the bacterium. The plasmid copy number in the cell was approximately 20 and the activity of the threonine operon enzymes was increased 40–50 times. The organism produced $30 \text{ g dm}^{-3}$ threonine compared with the $2-3 \text{ g dm}^{-3}$ of the non-manipulated strain. Miwa, Nakamori and Mimose (1981) applied *in vitro* recombination to the improvement of the threonine producer, *E. coli* $\beta 1M4$, which was resistant to $\alpha$-amino-$\beta$-hydroxy valeric acid (a threonine analogue) and auxotrophic for isoleucine, methionine, proline and thiamine and yielded $3-6 \text{ g dm}^{-3}$ threonine. The threonine operon from the production strain was inserted in the plasmid pBR322 and the hybrid plasmid introduced into a threonine auxotroph derived from the producer, $\beta 1M4$. The optimization of the cultural conditions of the recombinant strain resulted in a productivity of $65 \text{ g dm}^{-3}$ threonine.

The application of the techniques of genetic manipulation to the industrially important *Corynebacterium glutamicum* has been hindered by the lack of suitable vectors for gene cloning. However, two patents (Ajinomoto, 1983; Kyowa Hakko Kogyo, 1983) were filed in 1983 describing the isolation of corynebacterium plasmids which may be used as vectors.

The improvement of the production of commercially important enzymes has been achieved by the techniques of genetic manipulation. Enzyme yield may be increased by incorporating the chromosomal gene coding for the enzyme into a plasmid which may then be introduced into the original strain (or a different one) and maintained at a high copy number. Colson *et al.* (1981) cloned a *Bacillus coagulans* gene coding for a thermostable $\alpha$-amylase into *E. coli* where it was replicated and maintained at a high copy number with resultant high enzyme production. Penicillin acylase production in an *E. coli* strain has also been improved by introducing the relevant gene into a plasmid which was then incorporated into the original strain (Stoppok *et al.*, 1980).

The application of *in vitro* recombinant DNA technology to the improvement of secondary metabolite formation is not as advanced as it is in the primary metabolite field. The main reasons for this lack of progress is the lack of information on the basic genetics of secondary metabolite production coupled with the difficulties inherent in the study of the production of compounds that are not essential to the growth of the producing cells. However, considerable advances have been made in the genetic manipulation of the streptomycetes due to the pioneering work of Hopwood and his co-workers. Techniques have been developed using protoplasts as recipients for genetically modified streptomycete plasmids, three of which have been used to clone genes involved in antibiotic synthesis. Although the manipulated cultures did not produce higher levels of antibiotic they should assist in the understanding of the structure and control of secondary metabolism genes. Crameri, Davies and Thompson (1985) succeeded in improving the yield of aminoglycoside antibiotics from *Streptomyces kanamyceticus* and *Streptomyces fradiae* by cloning the gene for aminoglycoside resistance into the producing organisms. Thus, genetic manipulation techniques are giving exciting results in the field of secondary product formation, and several workers have suggested how these advances may be extended. Malik (1982) has suggested that the isolation of messenger RNA synthesized at the onset of secondary metabolism should provide templates for the synthesis of cDNA, and therefore the genes which may be coding specific enzymes of secondary metabolism. Kurth and Demain (1984) have advocated the use of 'shotgun' cloning either in the producer cells to obtain gene amplification, or interspecific which may lead to the synthesis of new antibiotics. A long-term aim would be to transfer the genes for secondary metabolism from the natural producer cell to a more 'convenient' organism for production purposes, such as *E. coli*.

## 9.4 Summary

Thus, the genetic improvement of industrial strains may be achieved by mutation and recombination. The techniques of mutant selection have resulted in the attainment of vast increases in productivity over a long time period. The use of traditional recombination techniques by industrial companies has very probably resulted in significant advances in fermentation performance. The advances made in *in vitro* recombinant DNA technology are already resulting in improved 'traditional' processes as well as the establishment of new ones. However, it must be remembered that the 'ideal' industrial strain is likely to be developed by the combination of all these techniques and that recombinant DNA technology cannot be pursued in isolation.

# SECTION IV

# Genetic Engineering

*S. A. Boffey*

## Assumed concepts

RNA and DNA structures; the genetic code; basic aspects of transcription and translation.
Protein structure.
Principles of enzyme action; affinity of enzyme for substrate.
Meanings of: gene, genome, homozygous, transformation, prokaryote, eukaryote.

*Useful books*
Lewin, B. (1983) Genes. Wiley, New York.
Mainwaring, W. I. P., Parish, J. H., Pickering, J. D. and Mann, N. H. (1982) Nucleic acid biochemistry and molecular biology. Blackwell, Oxford, U.K.
Stryer, L. (1981) Biochemistry, 2nd edn. Freeman, San Francisco.
Watson, J. D., Tooze, J. and Kurtz, D. T. (1983) Recombinant DNA—a short course. Scientific American Books. Freeman, New York.

# Chapter 10

# *The aims of genetic engineering*

## 10.1  Techniques of gene manipulation

Genetic engineering implies 'changing genes'. These changes may occur as the result of the transfer of a gene from its normal location into a cell which does not normally contain it; or a gene may have its sequence altered in some way, so that it is a different gene. Such an altered gene may, of course, also be transferred to a new cell type. To avoid confusion with conventional breeding techniques, or with the naturally occurring process of gene transfer by plasmids, it is assumed that genetic engineering, or genetic manipulation, also involves some processes which are carried out *in vitro*.

### 10.1.1  *Conventional breeding*

Why is there a need for genetic engineering? It is true that a great deal has been and will continue to be achieved by conventional breeding programmes. Methods used for the 'breeding' of improved microbial strains, using conjugation or, more rarely, parasexual gene transfer between strains, have been described in Section 9.3.1, and will not be repeated here. In general, the breeder aims to introduce a desirable gene into an organism which already has many attractive attributes. This is done by crossing the organism with another which contains the desired gene, but which is in many other respects less than perfect. The hope is that the progeny of such a cross will combine the good features of the first organism with the desired gene of the second. Unfortunately, the chances of such a perfect combination of genes occurring tend to be vanishingly small; in fact, there is just as much likelihood of obtaining progeny with only the undesirable features of each parent. Consequently, the breeder has to select those progeny which contain the desired gene, and which have also retained many of the good features of the first parent. If these are then crossed with the first parent, the resulting progeny will, on average, contain fewer unwanted genes than did the first generation, although some will also have lost the desired gene and must therefore be discarded. By such repeated back-crossing of selected progeny, the breeder can reach the stage where, for practical purposes, he has effectively transferred only the desired gene into the genome of the first organism. Even then, further crosses must take place between the progeny to generate organisms which are homozygous for the 'transferred' gene, and will therefore be 'true' breeding.

It should be obvious that conventional breeding takes a long time, even when tricks are used to stack the odds in the breeder's favour. The most rapid progress can be made when the organisms to be crossed are already very similar, since this will reduce the number of back-crosses needed. However, the most useful genes tend to be found in organisms which are very different from the 'target' organism. If crossing is possible, the production of a 'true' strain will be very time-consuming, but feasible. Unfortunately, fertilization can usually only be achieved between organisms of the same, or closely related, species, so the 'pool' of genes available to any organism via conventional breeding is often very limited. This is why it is so important to maintain breeding stocks of a wide variety of 'outclassed' varieties of plants and animals, since they act as the gene pool for further breeding programmes.

The variety of the gene pool available to a species can be increased by the creation of mutants, using chemical mutagens or high energy radiation, as has been described for micro-organisms (see Section 9.2). However, since there is no control over which genes are mutated, the vast majority of such mutations will be lethal, harmful, or neutral. It may be that a beneficial mutation is generated but not detected, owing to the simultaneous creation of a lethal mutation in another gene of the same cell. Clearly mutagenesis is only useful if highly efficient methods are available for screening to detect beneficial mutations (see Section 9.2). The method is also restricted to unicellular organisms or those which can be grown from cultured cells or embryos, since its random mode of action rules out the possibility of producing identical mutations in all the cells of an animal or plant by treatment of the whole organism. As will be seen, this limitation also applies to most types of genetic engineering.

### 10.1.2   *Protoplasts and cell cloning*

As described in Section 9.3.2, the efficiency of recombination between the genomes of different microbial strains can be greatly increased by the use of 'protoplast fusion'. This technique can also be applied to the cells of plants and animals. When the walls of plant cells are digested by pectinases and cellulases, in a medium which is iso-osmotic with the cell contents, spherical protoplasts are produced, being effectively naked cells bounded only by their plasma membranes. With careful handling these protoplasts will regenerate cell walls, will grow and divide, and each protoplast will eventually give rise to a cluster of cloned cells which can be grown indefinitely as undifferentiated callus tissue. In many cases whole plants can be produced from the callus by altering the levels of plant growth hormones in the medium, and so it is possible to produce a limitless number of plants cloned from a single protoplast.

Most importantly, by use of a reagent such as polyethylene glycol, or a pulsed electrical field, protoplasts can be induced to fuse with each other prior to cell wall regeneration (Shepard *et al.*, 1983). Such fusions can occur between cells of species which are sexually incompatible, resulting in the formation of hybrid cells. Unfortunately these hybrids are unstable, and one complete set of chromosomes is eventually eliminated, but not before there has been some transfer of genes between the two. Since the exchange of DNA is random, there is only a small chance that any fusion product will contain a specific 'foreign' gene, consequently some type of selection must be used to pick out the wanted cells. This can often only be done once plants have been regenerated from each cell line, unless the desired gene codes for a product which can be detected biochemically at the callus stage. In spite of its unpredictability and technical difficulty, protoplast fusion is a valuable addition to the plant breeders' more conventional techniques. For example, it has been used to transfer disease resistance from primitive *Solanum* species into modern commercial potato cultivars, with which they cannot be crossed sexually.

Animal cells also can be induced to fuse, and they normally form unstable hybrid cells; however, since whole animals cannot be regenerated from single cells, this technique has not been used as an adjunct to conventional animal breeding programmes. Animal cell fusion has, however, assumed enormous importance since the discovery by Milstein and Kohler that normal antibody-producing cells could be fused with malignant (myeloma) cells to produce 'hybridoma' cells, each of which would then produce a single type of antibody. the hybridoma cells can be separated from each other and then cultured indefinitely, each clone producing a unique product, known as a 'monoclonal' antibody. These antibodies are proving of immense value in medicine and molecular biology, since they are far more effective than mixtures of different antibodies in discriminating between similar antigens.

Another method of producing new characteristics in a plant is based on the phenomenon of 'somaclonal variation', in which it is found that plants cloned from different cells of the same original plant may possess different, heritable characteristics. It is not possible to predict what new characteristics could appear, and many will be undesirable, but the method has proved its value in the production of new, improved varieties of potato and wheat.

### 10.1.3  *Potential products of genetic engineering*
There is one important area of genetic manipulation in which conventional breeding is completely powerless, namely the transfer of a gene from a higher organism into a bacterial or yeast cell, with the aim of obtaining synthesis of the gene product by the transformed cells in culture. But, since the gene products are produced by the animals or plants from which they are derived, why should we go to all the trouble of persuading lower organisms to do the same job? The answer is usually based on the cost of isolating adequate quantities of product from the natural source, or on the sheer impossibility of obtaining sufficient starting material.

10.1.3.1  *Pharmaceuticals.* One has only to consider a selection of some of the polypeptides which are of pharmaceutical value to appreciate the enormous potential of genetic engineering for their production. For example, many haemophiliacs are unable to stop bleeding because of their inability to produce a polypeptide called human blood clotting factor VIII, which is normally present in human blood, though at very low concentrations. Fortunately, they can be prevented from bleeding to death by the injection of factor VIII, purified from vast quantities of human blood. Not only is the preparation of factor VIII a very expensive process, owing to its low initial concentration in a liquid which is in short supply and of value in its own right, but, since each preparation is derived from the blood of many thousands of donors, there is a significant risk that dangerous viruses, such as those for hepatitis or AIDS, may be transmitted to haemophiliacs in their injections of clotting factor.

Similar problems of production arise with other polypeptides which have to be extracted from human serum or tissues: tissue plasminogen activator could be used to dissolve unwanted blood clots blocking blood vessels, thus relieving thrombosis; calcitonin, possibly in conjunction with parathyroid hormone, may be of great value in stimulating a strengthening of the brittle bones so often found in elderly people. A nerve growth factor has been identified which could promote the repair and regeneration of damaged nerves. Human growth hormone is a polypeptide already used to treat children suffering from hypopituitarism, a condition that would otherwise result in severely stunted growth, or dwarfism. Fortunately, the relatively low incidence of dwarfism means that it can be controlled using hormone extracted from the pituitaries of human cadavers, but there are indications that human growth hormone could also be of value in the treatment of burns, to promote the healing of wounds and fractured bones, and to treat bones weakened by decalcification. Since supplies of hormone from human pituitaries are almost entirely used up in treating dwarfism, any expansion of the application of human growth hormone can only occur if it can be produced by genetically engineered micro-organisms.

Possibly the best known example of a genetically engineered polypeptide used for the treatment of humans is insulin. This hormone is essential for the correct regulation of blood sugar levels; if it is not produced by the pancreas, diabetes will result, which must then be kept under control by injections of insulin. Since there are estimated to be about 60 million diabetics in the world, of which about 25 million are in developed countries (that is, they can pay for their drugs), there is a huge market for insulin. For most diabetics, insulin obtained from pigs or cattle appears to work as effectively as the human variety, as might be expected in view of the fact that they differ from human insulin in only one or three amino acids respectively. However, some diabetics can develop an allergic response to the animal insulins, and there is debate over the long-term side effects of using a 'foreign' hormone. These reservations prompted attempts to produce 'human' insulin by genetically engineered bacteria, and it looked as if the first company to succeed would make a fortune. Unfortunately for these companies, research into ways of modifying insulin chemically was making headway, to such an extent that it is now possible to modify pig insulin to exactly the same structure as human insulin. The battle between the two processes for insulin production will therefore be fought on economic grounds and, in view of the plentiful supply of pig pancreases, victory for the biotechnologists seems unlikely.

Another group of polypeptides which has caused a lot of excited speculation is the interferons. These molecules are produced naturally by the human body in response to viral infections, and appear to have antiviral and antitumour properties. However, as with many such compounds, so little could be produced from human tissue (white blood cells and fibroblasts in the case of interferons) that it was not possible to carry out satisfactory clinical trials to establish their properties. In view of their *possible* therapeutic effects, programmes were established with the aim of producing interferons on a large scale in bacteria. Three types of interferon have been produced in this way. Although they certainly do not appear to be the 'wonder drugs' that some had predicted, they do seem to have a significant effect on the immune system, resulting in greatly improved resistance to a range of viruses, and they are also effective in treating a few types of cancer, notably a rare leukaemia called hairy cell leukaemia. Ironically, they may be of most use against cancer by revealing more about how the immune system can detect and eliminate tumours.

Successors to interferon, as the most interesting 'anticancer' drugs, may be tumour necrosis factor (TNF) and lymphotoxin (LT). These polypeptides are produced by human cells, and they are able both to recognize tumour cells and to destroy them selectively. Although they are different molecules, they have some quite extensive sequences in common, and two regions which are highly conserved (that is, likely to have the same three-dimensional structure in both molecules) could turn out to be responsible for tumour cell recognition and cell destruction. The genes for both TNF and LT have been cloned, and clinical trials of the bacterially synthesized molecules should reveal how effective they are against tumours and, most importantly, if they produce harmful side effects (Gray *et al.*, 1984; Pennica *et al.*, 1984).

10.1.3.2 *Vaccines.* Medical applications of genetically engineered micro-organisms are not limited to the synthesis of human polypeptides. A very attractive use of such systems is in the production of vaccines. The idea behind a vaccine is that it should contain molecules which produce the same antigenic response as the virus to which it will give immunity, without causing the harmful proliferation of viral particles within the body. This can be achieved by the use of virus particles which have been in some way made harmless, but which still have the same outward appearance. There are two problems, however. Incomplete inactivation of the virus may allow a few 'live' particles to survive, which would therefore actually be a source of infection; some occurrences of foot and mouth disease have been attributed to such vaccine preparations. The other problem is that the virus has to be grown in animals, and in certain cases it will grow only in humans; in such cases human cell cultures must be used, which are expensive to maintain, and very slow growing by comparison with bacterial cells. Since the market for vaccines can be immense (200 million people worldwide suffer from hepatitis B; 800 million doses of foot-and-mouth disease vaccine are given to livestock each year), it has been economically worthwhile to live with these problems. However, it is the viral coat proteins which are antigenic, and so preparations of pure coat protein can act as an effective vaccine. This immediately opens up the possibility of cloning and expressing the genes for the coat proteins in bacteria, and using the bacterial product as a vaccine. Such vaccines would be guaranteed free of live virus, and should be cheaper than those produced using animal cell cultures.

10.1.3.3 *Enzymes*. Enzymes can only be used on an industrial scale if their production costs are low (see Sections 13.2, 14.3 and 14.6.2), and this effectively means that they must be derived from cultures of micro-organisms. In view of the immense variety of reactions carried out by microbes, it is likely that any enzyme which will be needed by biochemical engineers is already being made by some micro-organism. However, it is also likely that a certain amount of genetic manipulation would improve the yield of enzyme, perhaps by transferring its gene into a faster growing, safer, more efficient host, or by attaching the gene to a stronger promoter in order to increase its level of expression. By attaching suitable sequences to the gene it may be possible to trick the cell into exporting the enzyme, thus making its isolation easier.

Many enzymes become unstable when isolated from the cell, making them unsuitable for industrial use. Genetic engineering offers the possibility of modifying the structure of an enzyme to increase its stability, alter its pH or temperature optima, change its affinity for substrate, etc., but such 'tuning' of enzymes will depend on a vast improvement in our understanding of exactly how enzymes work.

10.1.3.4 *Improved cell lines*. Cell cultures are grown either for the production of a valuable molecule, such as an antibiotic, or in order to generate high quality biomass from a cheap feedstock, as in the case of single cell protein production. In both cases there are strong commercial reasons for using cells which grow as efficiently as possible, in terms of energy conversion, and which are as well suited to their particular function as possible. Several examples of the use of genetic engineering for enhancing the efficiency of growth of micro-organisms, or for increasing their yields of a useful product are presented in Section 9.3.3.

10.1.3.5 *Animals*. To the non-scientist, the term 'genetic engineering' probably conjures up visions of a modern Dr Frankenstein creating monstrous new creatures in his laboratory. Ironically, the genetic manipulation of animal genomes seems unlikely to make as much progress over the next few years as that of other organisms. As will be seen, genetic manipulation involves the selection of those cells whose genomes have been correctly altered from those which are either unaltered or incorrectly modified, followed by proliferation of the selected cells ('cloning'). This is the end of the story when we are dealing with unicellular organisms, such as bacteria or yeast, or with animal or plant cell cultures; but, for the production of genetically engineered animals or plants, whole organisms must then be regenerated from the cloned cells. At present we are unable to regenerate animals from cloned cells, except to the limited extent of subdividing clusters of cells at the early stages of embryo development. Assuming this obstacle can be removed, there is still the problem of knowing which genes to modify. The genomes of animals are immensely more complex than those of bacteria, yet their genetic analysis is made difficult by the relatively small populations available for investigation, and by their long generation times. Add to this the knowledge that characteristics such as growth rate, milk yield, disease resistance and shape are determined by interactions between many different genes, and it becomes evident that we still have a long way to go before the quality of livestock can be improved by genetic engineering. However, in spite of these obstacles, there have already been some successful attempts to obtain expression of foreign genes in whole animals (see Section 11.3.3).

10.1.3.6   *Plants*. By contrast, the genetics of the major crop plants are very well characterized. The locations of a wide range of genes, of known function, have been determined, and many of these genes have been isolated, cloned, and even sequenced. Such detailed analysis has been possible owing to the ease of producing vast numbers of plants from crossing experiments. The problems of gene interaction to determine plant characteristics remains, but it has often been possible to identify one or two genes that have an especially strong influence on a particular characteristic. Consequently there are many plant genes, coding for such attributes as resistance to disease, stress, or herbicides, and for seed storage proteins or components of the photosynthetic system, which the breeder might wish to transfer from one species to another. It is possible to clone the cells of most plants by growth as a callus or in suspension culture, and whole plants can be regenerated from many of these. In such cases there appears to be no reason why genetic engineering should not be used as a faster alternative to conventional breeding, and to bypass any barriers which prevent crosses between plants of different species. Unfortunately, it is still not possible to regenerate plants from cell cultures of the most important, monocotyledonous (grass-like) crops, such as wheat or maize, and this remains the most serious impediment to their genetic manipulation (Ozias-Akins and Lorz, 1984).

The plant genetic engineer's fantasies do not stop at the transfer of genes from one plant to another; they range as far as introducing microbial genes into plants. The most attractive genes include those which code for enzymes responsible for the degradation of certain herbicides, and which might therefore endow a plant with resistance to them. Such resistance would be very useful in a crop, allowing herbicides to be used selectively against weeds *after* emergence of the crop. Other potentially valuable genes are those which result in the production of insecticidal polypeptides by some bacteria, such as *Bacillus thuringiensis*.

Probably the most ambitious proposal is to introduce the genes for nitrogen fixation into plants from an organism such as *Klebsiella pneumoniae*. This would involve the transfer of a large complex of 17 genes, known as the *nif* gene cluster, and a certain amount of manipulation of the cluster would be needed to ensure gene expression. Because of the prokaryotic origin of the *nif* cluster, it has been suggested that the most sensible strategy would be to transfer it to the chloroplast, whose transcriptional and translational machinery has many prokaryotic features. This approach also has the advantage that chloroplasts are geared to the production of ATP and reducing power, both of which are needed for nitrogen fixation (Merrick and Dixon, 1984). There are, of course, problems, not the least of these being the sensitivity of nitrogenase to the oxygen which chloroplasts produce, and also the absence of a method for transforming chloroplasts. However, in view of the massive annual consumption of artificial fertilizers, there is every incentive to overcome these problems. An alternative approach, which is receiving considerable attention, is to introduce into crop plants genes responsible for association with nitrogen-fixing micro-organisms, using genetic engineering.

It is easy to be carried away by the exciting possibilities of genetic manipulation, and to assume that the ideas discussed above will come to fruition within the next few years. There are enormous commercial pressures for rapid progress, but also many technical problems, and some serious gaps in our knowledge of the processes which we would like to alter.

## 10.2  Summary

Genetic engineering is defined as the changing of genes, using *in vitro* processes. This is contrasted with conventional breeding processes, in which genes are transferred by sexual or conjugative means. Since such gene transfer is a relatively random process, the breeder must select for desirable characteristics, and must rely on mutations to generate 'new' genes. A serious limitation of this approach is that it is not possible to transfer a gene between organisms that are totally unrelated to each other.

By contrast, genetic engineering is especially useful when one wishes to introduce a gene from a higher organism into bacteria or yeast. Such manipulations can result in the microbial synthesis of pharmacologically active polypeptides, vaccines, or enzymes for industrial use. They can be used for the improvement of industrially important microbial cell properties, or for the genetic modification of animals. However, in the latter case there is the major problem of modifying all cells within the organism, since animals cannot be regenerated from single, cultured cells. Plants, on the other hand, can be more readily manipulated, and there are real prospects of introducing resistance to diseases and herbicides, increasing yield, directing the synthesis of insecticidal polypeptides, and introducing nitrogen fixation into non-leguminous crops.

# Chapter 11

## *Techniques of genetic engineering*

### 11.1 Outline of gene cloning

Gene cloning was initially made possible by technical developments, such as the isolation of enzymes which would cut DNA at precisely defined locations (restriction endonucleases), or those which would covalently join DNA fragments (ligases), and advances still often depend on the development of new enzymes or other biochemicals. We are experiencing an explosion in the publication of new techniques related to cloning, and this proliferation of technical information, combined with an excess of jargon, can easily blind the novice to the basic principles of cloning. Once these are understood, it is much easier to see where new techniques fit in to the cloning process. The aim of this section is to outline the strategies employed for gene cloning, and to introduce those techniques which are most commonly used.

There are several possible reasons for wishing to clone a particular piece of DNA. One might wish to introduce into bacterial cells a gene coding for a valuable polypeptide, in such a way that the cells would produce large amounts of the polypeptide. This is the approach which has received most publicity, resulting in the production by bacteria of such polypeptides as insulin, interferon and growth hormone. As will be seen later, DNA can also be cloned in eukaryotic cells, with the aim either of using the cells as biochemical 'factories', or in order to modify the properties of the cells to make them better suited to our needs. However, in practice gene cloning is carried out most often to produce enough DNA for subsequent analysis by, for example, sequencing.

Gene cloning is essentially the insertion of a specific piece of 'foreign' DNA into a cell, in such a way that the inserted DNA is replicated and handed on to daughter cells during cell division. This process occurs naturally, as demonstrated by the rapid spread of multiple resistance to antibiotics through populations of bacteria under the appropriate selective pressure. However, transfer of DNA normally occurs only between closely related strains of bacteria, and there is little control over which DNA fragments are transferred. So natural DNA transfer, or transduction, is too unpredictable and limited in range to be useful for gene cloning.

The main steps of gene cloning are summarized in Fig. 11.1, and involve the following procedures:

(1) Isolation of the gene (or other piece of DNA) to be cloned.
(2) Insertion of the gene into another piece of DNA, called a vector, which will allow it to be taken up by bacteria and replicated within them as the cells grow and divide. Although Fig. 11.1 shows the use of a plasmid vector, since this is still a widely used system, other types of vector can also be used, such as viral DNA or cosmids.
(3) Transfer of the recombinant vectors into bacterial cells, either by transformation or by infection using viruses.
(4) Selection of those cells which contain the desired recombinant vectors.
(5) Growth of the bacteria; this can be continued indefinitely, to give as much cloned DNA as is needed.

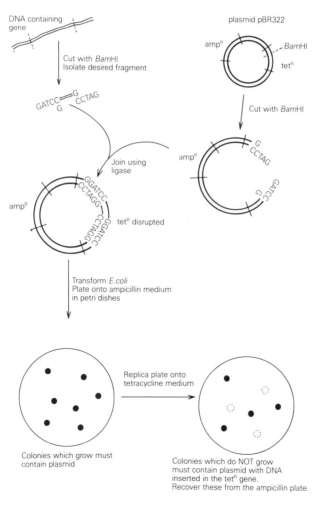

**Fig. 11.1** Outline of gene cloning

(6) The vector, containing its inserted DNA, is normally recovered from the cells at this stage, and the insert cut out and separated from vector molecules. However, sometimes the aim is to obtain expression of the gene within the bacterial cells, especially if the gene codes for a polypeptide which is difficult, or expensive, to prepare by other means.

## 11.2 Gene cloning procedures

### 11.2.1 *Restriction endonucleases*

At the heart of gene cloning lie the restriction endonucleases. These enzymes, which occur naturally in bacteria as a weapon against invading viruses, will cut both strands of DNA molecules whenever certain nucleotide sequences occur. The class of enzymes most used for gene cloning is known as type II; each of these recognizes a particular sequence of nucleotides, usually four or six nucleotides long, and normally cuts the DNA strands within this sequence. The cuts can be staggered, producing short single-stranded projections, or they can be opposite each other, in which case 'blunt' ends will result. Figure 11.2 illustrates the types of sequence which can be recognized by restriction enzymes, and shows where cuts are made. Note that those sequences which are cut internally (the majority) are symmetrical; that is, when read from 5' to 3', both strands have the same sequence. Staggered cuts will yield single stranded ends that are identical and complementary to each other. Consequently, at low temperatures there is a tendency for the single strands to associate by base-pairing, causing a reversible linking together of fragments which have been cut by the same enzyme; such fragments are said to have 'sticky' or 'cohesive' ends. There is now a huge range of restriction endonucleases available commercially, so that blunt or sticky ends can be produced from most tetranucleotide and hexanucleotide sequences.

The isolation of genes for cloning, insertion of DNA into vectors, recovery of inserts from cloned vectors, and construction of new vectors all depend on our ability to cut

| Enzyme | Recognition sequence | Products | | Conventional representation |
|---|---|---|---|---|
| EcoRI | 5'—GAATTC—3'<br>3'  CTTAAG—5' | 5'—G<br>3'—CTTAA | AATTC—3'<br>G—5' | GAATTC |
| BamHI | 5'—GGATCC—3'<br>3'—CCTAGG—5' | 5'—G<br>3'—CCTAG | GATCC—3'<br>G—5' | GGATCC |
| HpaI | 5'—GTTAAC—3'<br>3'—CAATTG—5' | 5'—GTT<br>3'—CAA | AAC—3'<br>TTG—5' | GTTAAC |
| Sau3A | 5'—GATC—3'<br>3'—CTAG—5' | 5'—<br>3'—CTAG | GATC—3'<br>—5' | GATC |

**Fig. 11.2** Recognition sequences of some restriction endonucleases. Arrows indicate positions of cleavage. Note that the cohesive ends produced by *Sau*3A are complementary to those produced by *Bam*HI, yet the former recognizes a tetranucleotide sequence while the latter recognizes a hexanucleotide

DNA at well-defined sites—hence the central importance of restriction endonucleases in gene cloning.

### 11.2.2 *Isolation of DNA to be cloned*

Depending on the aims of the experimenter, the DNA can come from a variety of sources. Perhaps the intention is to isolate and clone a particular gene for subsequent sequencing, or for investigation of its transcriptional control. Inevitably the gene will form only a small part of the entire genome (approximately 0.03% of the *Escherichia coli* genome and only $0.03 \times 10^{-3}\%$ of mammalian genomes), so a way must be found to identify the gene and to 'pull it out' from the other parts of the genome either before or after cloning. This job is made much easier if the corresponding mRNA is available in a fairly pure form. If the protein for which it codes is a major product of the cells, then the mRNA will already be enriched for the desired species. However, procedures such as size fractionation of mRNA, *in vitro* translation of the fractions, precipitation of translation products with specific antibodies, etc., will be needed for the enrichment of those mRNAs (the majority) which are less abundant. Methods of obtaining specific mRNAs are described in Section 11.2.6, and it is worth noting that this can be the most difficult step in cloning, particularly if the gene product is not well characterized.

Such purified mRNA can be used to direct the synthesis of a complementary DNA (cDNA) strand, using the enzyme reverse transcriptase (Fig. 11.3); this results in a

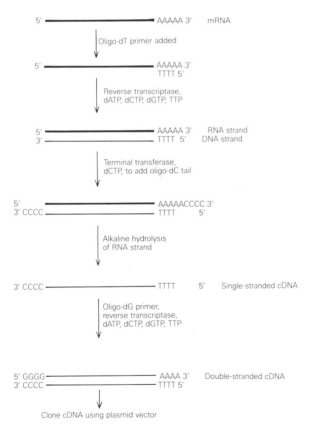

**Fig. 11.3** Synthesis of cDNA

DNA/RNA hybrid. After removal of the RNA strand using alkali, the single-stranded DNA can act as a template for the synthesis of a complementary DNA strand in the presence of DNA polymerase. One might expect the resulting double-stranded cDNA to be identical to the gene which coded for its mRNA, and this is so in the case of prokaryotic genes. However, the majority of eukaryotic genes contain intervening sequences (introns) which interrupt their coding sequences. The whole gene, including introns, is transcribed, the introns are cut out of the mRNA, and the remaining pieces (exons) are spliced together to give the mature mRNA which is translated, and from which cDNA is made. Since introns are not present in functional mRNA, cDNA can be used in place of the true gene when the aim of cloning is gene expression. Clearly, however, this approach cannot be used if one wishes to sequence the native gene, including its introns. In the latter case, cDNA can be radioactively labelled by nick translation, and used as a hybridization probe for the identification of DNA fragments containing the gene.

A procedure commonly employed when dealing with a poorly characterized genome is the formation of a 'gene library', in which the entire genome is cut randomly into large pieces, about 10 kilobases (kb) or more in size, using partial digestion with several restriction enzymes, and these pieces are all cloned, without any attempt to select for particular sequences. When the bacteria containing these fragments are plated out to give rise to individual colonies, each colony will contain a sample of the genome, and there is a finite chance that the desired gene will be present in any one sample. Provided enough colonies are produced, one can be almost certain that at least one will contain the gene. The larger the genome and the smaller the average fragment size, the more colonies will have to be grown to be sure of finding any particular gene. Calculations show that, for 10 kb fragments, about $1.5 \times 10^3$ colonies are needed for the *E. coli* genome, and nearly $2 \times 10^6$ for *Homo sapiens*. In the latter case the number of colonies required is clearly unmanageable, and therefore much larger fragments, about 40 kb in size, must be cloned. As will be described in Section 11.2.7, such large fragments need special vectors for cloning. We shall look at methods of identifying colonies which contain specific genes later (see Section 11.2.6).

Nucleic acid chemistry has advanced so rapidly that it is now a routine procedure in many laboratories to synthesize long DNA molecules of any desired nucleotide sequence. In fact it is possible to buy machines which will carry out each step of the synthesis automatically; all the operator does is to supply reagents and type in the sequence of nucleotides to be produced. Consequently, if the amino acid sequence of a polypeptide is known, it may be easier to synthesize a 'gene' which codes for that sequence than to try and isolate the natural gene; such an approach was used for the bacterial synthesis of the polypeptide hormone somatostatin.

### 11.2.3 *Plasmid cloning vectors*

In order to clone a fragment of DNA it is not sufficient to transfer it into a bacterial cell, in the hope that it will be replicated along with the cell's own DNA. Unless the foreign DNA contains a sequence of nucleotides which is recognized by the host bacterium as an 'origin of replication' it will not be replicated. So it is usually necessary to attach the DNA to another fragment which contains an origin of replication. Such origins of replication occur naturally in plasmids, which are small, circular molecules of DNA, found in many bacteria. They are separate from the main chromosomal DNA, yet are replicated and usually passed on to daughter cells during cell division. Some very important genes can be carried by these plasmids, including those for resistance to antibiotics, for toxin or antibiotic production, for nitrogen fixation, and for enzymes needed for the degradation of a large number of 'unusual' substrates such as herbicides or industrial effluents. Only a few of these genes are present in any one plasmid but, owing to the frequent exchange of genetic material between plasmids and chromosomal DNA, new collections of genes are continually emerging, and so help bacterial populations to respond quickly to new environmental conditions. This can be a serious problem in hospitals if a range of antibiotics is in use, since gene transfer via plasmids can soon result in the appearance of bacteria with resistance to several different antibiotics at once.

For gene cloning, plasmids offer a convenient source of replication origins; but most of their genes are superfluous, and would make the molecules difficult to manipulate. Therefore plasmids have been constructed from naturally occurring forms, preserving only those features which will aid cloning. One of the most widely used plasmids, pBR322, was constructed by excising parts of naturally occurring plasmids, using restriction enzymes, and rejoining them in the correct orientations, in a series of manipulations which makes solving Rubik's cube look simple. The resulting plasmid demonstrates many of the features which are desirable in a plasmid cloning vector (Fig. 11.4). The molecule contains an origin of replication, which was derived from a plasmid related to the naturally occurring plasmid Col El. This origin is particularly useful because it is 'relaxed', that is, its replication is not linked to that of the chromosomal DNA, and hence initiation of plasmid replication can be more frequent than that of the main bacterial DNA. Consequently multiple copies of the plasmid accumulate in each cell. This process can be taken still further if chloramphenicol (an inhibitor of protein synthesis) is added to the cell culture, since protein synthesis is needed for chromosomal, but not plasmid, replication. In this way chloramphenicol can be used to 'amplify' the plasmid, resulting in up to 3000 copies per cell.

The plasmid has also been carefully constructed so that it contains unique recognition sites for 20 different restriction endonucleases. This means that any one of the enzymes will make a single cut in the circular plasmid, generating a single, linear

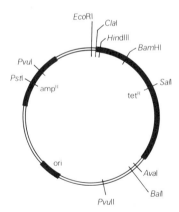

**Fig. 11.4**   Map of the plasmid pBR322. Regions coding for resistance to ampicillin (amp^R) and tetracycline (tet^R), and for the origin of DNA replication (ori) are indicated by solid blocks. The cleavage sites of a selection of restriction endonucleases are indicated, some of which cut the plasmid within an antibiotic resistance gene. Each of the enzymes shown cuts the plasmid at a unique site, and hence produces a single, linear molecule from the circular plasmid

molecule to which can be attached a DNA fragment for cloning; then the molecule can be reclosed to regenerate a slightly enlarged, circular plasmid. It will be seen from the examples in Fig. 11.4 that the sites of restriction are distributed around the plasmid, and this allows great flexibility in the approach to selecting for recombinant molecules, as described below.

Two sets of genes have been preserved in pBR322; these code for resistance to ampicillin and to tetracycline, and are very useful markers. The presence of plasmid in a bacterium will confer resistance to these antibiotics, and so such cells can be selected for by growth on medium containing either ampicillin or tetracycline. Furthermore, because some of the restriction sites are situated on the antibiotic resistance genes, it can be arranged that insertion of foreign DNA will inactivate one or other of the genes. For example, if the DNA is inserted at the *Bam*HI site, the tetracycline resistance will be destroyed; so recombinant plasmids will allow cells to grow in the presence of ampicillin, but will not protect them against tetracycline.

This provides the basis for a method of discriminating between those cells which simply contain plasmid, and those which contain plasmid with a DNA insert (see Fig. 11.1). Cells are plated out at low concentration onto a growth medium, solidified using agar, which contains ampicillin. After incubation, all cells containing a plasmid will have grown to form colonies. It would be possible at this stage to discover which cells contained plasmids with inserts, by using cells from each colony to inoculate a fresh plate of medium containing tetracycline instead of ampicillin. Cells which failed to grow on this plate would probably contain plasmids with inserts, and clones of these cells could then be recovered from the appropriate colony on the first plate. In practice, the easiest way of doing this screening is by 'replica plating', in which a sterile velvet pad is pressed onto the first plate of colonies, picking up some cells from each colony. The second plate is then inoculated by being pressed against the velvet pad. After incubation, colonies should be seen on the second plate in identical positions to those on the first plate, but there will be some gaps. It is then quite easy to spot which colonies have not grown on tetracycline, and these can be recovered from the first plate and grown up for further characterization of the inserts contained in their plasmids.

The plasmids used as vectors for cloning have been whittled down to the smallest possible sizes. This has been done for two reasons: first, small molecules of DNA are far less easily damaged by shearing during isolation than are large molecules. This increased stability allows relatively violent methods to be used for such processes as the deproteinization of plasmid preparations, and so makes it easier to obtain good yields of pure DNA. Secondly, small molecules are more efficiently taken up by bacteria during the process of 'transformation', an important consideration when dealing with very small quantities of valuable DNA. Plasmid size must therefore be kept to a minimum, and so it is obvious that the insertion of large fragments of DNA into a plasmid will result in a molecule which is physically unstable, and which may have an impracticably low efficiency of transformation. Such limitations on the maximum size of insert can cause problems, as, for example, when a gene library is being prepared from a eukaryotic organism. In such a case, it would be necessary to produce about $2 \times 10^6$ colonies of cells to be reasonably sure that at least one would have a plasmid carrying any particular gene as its insert, assuming an average fragment size of 10 kb. However, if fragments 40 kb in size could be cloned, only $7 \times 10^4$ colonies would need to be screened. Although this may still seem a large number, it is not, in fact, too difficult to deal with.

### 11.2.4 *Joining DNA*

The process of DNA cloning depends not only on the availability of restriction enzymes and vectors, but also on our ability to join lengths of DNA together covalently. If the DNA fragments to be joined have been cut using the same restriction enzyme, to give identical cohesive ends, they will tend to stick together if they collide, held by the hydrogen bonds between complementary base pairs. However, this association is only temporary, owing to the weakness of the bonds and the small number of bases involved. The linkage can be made permanent by using a 'ligase' enzyme, which will form a phosphodiester linkage between free 5'-phosphoryl and 3'-hydroxyl groups, thus joining two DNA molecules together, or circularizing a single, linear molecule. The main source of ligase is T4 phage, and this enzyme requires ATP to drive the ligation. In order to stabilize the base pairing of cohesive ends, ligation is usually carried out at temperatures between 4 and 10 °C, with long incubation times to allow for the low rate of enzyme activity at these temperatures. An important feature of T4 DNA ligase is its ability to join blunt ends of DNA together, provided the concentrations of enzyme and DNA are high enough.

It will be appreciated that the process of ligation, as described above, depends on the random collisions of DNA molecules to determine which fragments are joined together. Thus the products of a mixture of DNA fragments and linear vector will include not only vector with a single piece of inserted DNA, but also recircularized vector without any insert, lengths of DNA formed from two or more fragments, circularized fragments, two or more vector molecules linked together, etc. By careful choice of the concentrations of vector and 'insert' DNAs, the formation of plasmid containing a single insert can be favoured, but unwanted products cannot be eliminated completely.

For some purposes this type of ligation is adequate, especially when subsequent steps will select for the 'recombinant' product. However, there are ways of exerting more control over the products which can be formed. For example, if the vector is treated with alkaline phosphatase after restriction, its 5'-phosphoryl groups will be removed (Fig. 11.5). Ligase will only join 3' and 5' ends of DNA together if the 5' end is phosphorylated; and so, on ligation with untreated DNA fragments, links can only be formed between fragment and vector or fragment and fragment, thus preventing simple recircularization of the vector and increasing the yield of recombinant molecules. Of course, the fragment–vector linkage can only be made through one strand of the DNA, but this is sufficient to hold the molecule together, and the 'nicks' will eventually be repaired by bacterial host cells after transformation.

Ideally, both vector and 'insert' DNA should be cut with the same restriction enzyme, or with a pair which generate identical protruding ends (isoschizomers), but this is not always possible. In such cases, any cohesive ends can be converted to the blunt form, either by removing protruding nucleotides using S1 nuclease (which degrades single stranded DNA), or by filling in the single-stranded ends using DNA polymerase and all four nucleoside triphosphates. In principle, the blunt-ended molecules could then be linked using T4 DNA ligase. However, this is not usually done, since the resulting recombinant molecules would probably not contain sites for restriction at the points where the links have been made, and so it would be difficult to recover the inserted DNA from the vector after cloning. Consequently, it is far more useful to carry out blunt ended ligation between DNA fragments and 'linkers', which are chemically synthesized oligonucleotides containing one or more restriction sites

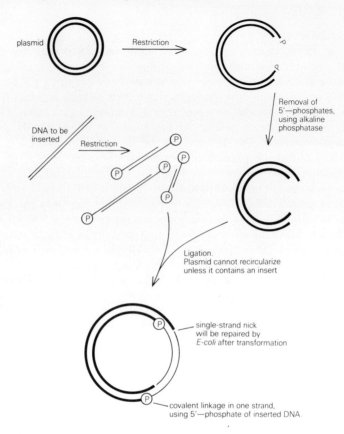

**Fig. 11.5**   Use of alkaline phosphatase to increase yield of recombinant molecules

(Fig. 11.6). If the vector contains, for example, a *Hind*III cleavage site in one of its antibiotic resistance genes, then the DNA fragments, after conversion to blunt-ended forms, would be ligated to linkers containing a *Hind*III site. Subsequent restriction of the fragments-plus-linkers using *Hind*III would generate molecules which could be ligated to the vector. There are so many linkers (and related molecules known as adaptors) commercially available that it is now possible to overcome almost all problems of incompatibility between vector and insert.

A technique known as 'homopolymer tailing' can be used to join blunt-ended molecules, and has the advantage that only intermolecular bonds, between vector and insert, can occur. Vector and insert are treated separately with terminal transferase and either dATP or TTP, so that poly dA tails are built up on the 3′-termini of one, and poly T tails on the other. On mixing, the complementary tails will result in stable, hybrid molecules which can be used for transformation. The disadvantage of this method is that it does not automatically create restriction sites on either side of the inserted DNA, and so recovery of inserts may be difficult. When dGTP and dCTP are used similarly with molecules which have been cut with the restriction enzyme *Pst*I, a *Pst*I cleavage site will be regenerated on ligation, and can be used to recover insert from the plasmid after cloning.

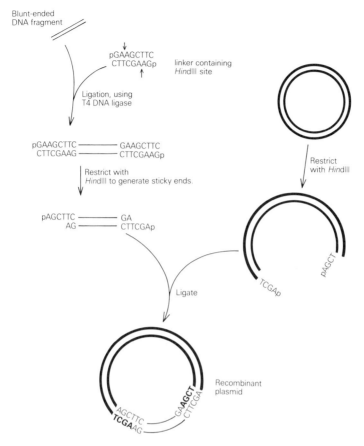

**Fig. 11.6** Use of linkers to join blunt-ended molecules

### 11.2.5  *Transformation and growth of cells*

Before the recombinant DNA can be bulked up by cloning it must be taken up by a suitable bacterial host cell, which is then said to be 'transformed', and which is usually a strain of *E. coli* lacking the restriction system which would normally degrade foreign DNA. Untreated cells will not take up DNA to any significant extent, and so they must be pretreated to make them 'competent'. This pretreatment usually involves incubation of exponentially growing cells with $CaCl_2$ at low temperatures, after which the DNA is added; a mild heat shock then results in uptake of the DNA. The efficiency of transformation is never high, and is influenced by the cell strain and the size and form of the DNA. Between $10^5$ and $10^7$ transformants can be obtained per microgram ($\mu$g) of supercoiled pBR322, but this rate drops to less than $10^4/\mu$g of religated plasmid, and decreases progressively as the size of inserted DNA increases. Linear DNA is almost completely ineffective in transformation. Even at its most efficient, transformation takes place with only about 0.01% of the DNA molecules. Since the process favours small molecules, it can present a problem when attempts are made to clone large fragments of DNA inserted in plasmids, and steps should be taken to prevent the recircularization of non-recombinant plasmid, using alkaline phosphatase (see Section 11.2.4).

The selection of transformed cells usually depends on their resistance to an antibiotic, and so it is important to incubate the cells in a medium without antibiotic for about an hour, to allow the plasmid antibiotic resistance genes to be expressed. The cells can then be plated out on a solid medium containing antibiotic, for selection of colonies containing recombinant DNA, as described in Section 11.2.3. Methods for the selection of specific colonies will be described below. Sometimes the aim is to clone a single species of plasmid (recombinant or otherwise) which has already been purified, and in this case there is no need to plate out the transformed cells prior to growth in a selective medium. See Chapters 6 and 7 for further information about transformation and bacterial cell growth.

### 11.2.6   *Selection of clones*

From the large number of colonies produced by transformation, how is it possible to pick out those few (perhaps only one in several thousand) which contain a particular fragment of DNA? One of the most useful methods is known as 'colony hybridization' (Fig. 11.7), and depends on the availability of a radioactively labelled 'probe', which is nucleic acid with a sequence complementary to at least part of the desired DNA. In this technique, the colonies of transformed cells are replica plated (see Section 11.2.3) onto a nitrocellulose filter placed on the surface of a gelled nutrient medium, and are incubated to form colonies on the filter (nutrients are able to diffuse through the filter to the cells). The cells are then lysed by treatment of the filter with alkali, and this also

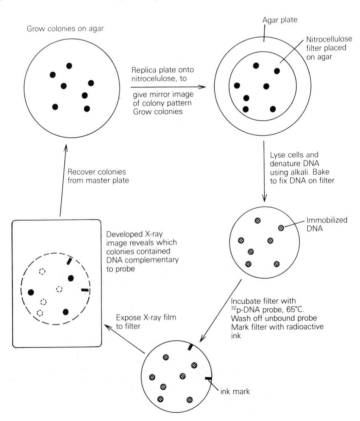

**Fig. 11.7**   Colony hybridization

denatures their DNA, which is fixed to the filter by baking. In this way, the pattern of colonies is replaced by an identical pattern of bound, denatured DNA. The filter is incubated with a solution of the labelled probe, which will hybridize to any bound DNA which contains sequences complementary to the probe; if this is carried out at temperatures of around 65 °C, hybridization will only occur when there is an almost perfect match between the two nucleic acids. Following thorough washing to remove unbound probe, the hybridized probe is located by autoradiography of the filter. It is then possible to say which colonies contain DNA complementary to the probe, and to pick these off the master plate for further growth and analysis.

Clearly colony hybridization relies on the availability of considerable amounts of a highly radioactive, specific probe. When the DNA of interest is a gene, its RNA transcription product is an obvious starting point for the production of a probe. This is one of the reasons why the genes for rRNAs and tRNAs are usually among the first to be mapped and isolated; the RNA products are plentiful and relatively easy to isolate, so they can be end-labelled for use as probes, or can be used as templates for the synthesis of radioactive cDNA probes (see Section 11.2.2). It is much harder to isolate a specific mRNA, especially if, like the majority, it is present in low abundance. Messenger RNA as a class can be separated from other types of RNA by virtue of the presence of a long poly-A tail on the 3′ end of almost all nuclear encoded mRNAs; hence mRNA will bind to an oligo-dT cellulose affinity column, from which it can be subsequently eluted (Fig. 11.8). This mRNA preparation can be enriched for the desired molecule by fractionation according to size, using sucrose density gradient centrifugation. But how does one detect which fraction contains the specific mRNA? Each mRNA fraction can be translated, using an *in vitro* translation system, based on rabbit reticulocyte lysates or wheat germ extract, and the presence of the desired mRNA is indicated if any of the translation products can be precipitated by antibodies to the protein for which the 'target' mRNA codes. If, as is often the case, it has not been possible to purify the protein product for antibody production, other features, such as enzyme activity or activity in a bioassay must be used to detect the desired translation product.

A very useful enrichment of the mRNA can often be obtained by precipitation of polysomes by antibodies to its purified protein, or by binding of the polysome–antibody complex to a protein A–Sepharose column. The greatest enrichments are produced by the use of monoclonal antibodies.

Having obtained an mRNA preparation enriched for a specific species, it is usually not worth attempting further purification of the RNA; instead, the mixture of RNAs is used to direct the synthesis of their corresponding cDNAs, and these are then cloned. Then begins the laborious task of identifying those clones which contain DNA complementary to the 'target' mRNA (Fig. 11.8). DNA from each clone is denatured and immobilized, and is then mixed with total mRNA under hybridizing conditions. After washing to remove unhybridized molecules, bound mRNA is released and translated *in vitro*, and the product analysed for production of the desired protein by SDS gel electrophoresis, enzyme assay, bioassay or immunoprecipitation. This technique is known as 'hybrid-release translation'. A related technique, though less useful for very low abundance mRNAs, is 'hybrid-arrested translation', in which the mRNA which is *not* bound to the immobilized DNA is translated, and one looks for the *absence* of a specific product. In practice, groups of colonies are pooled initially and, when a group gives a positive result, the clones making up that group are analysed individually.

With the advent of cheap and rapid methods for the synthesis of quite long

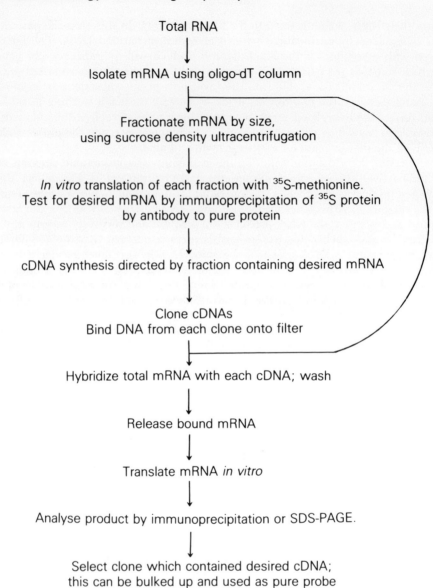

Total RNA

↓

Isolate mRNA using oligo-dT column

↓

Fractionate mRNA by size,
using sucrose density ultracentrifugation

↓

*In vitro* translation of each fraction with $^{35}$S-methionine.
Test for desired mRNA by immunoprecipitation of $^{35}$S protein
by antibody to pure protein

↓

cDNA synthesis directed by fraction containing desired mRNA

↓

Clone cDNAs
Bind DNA from each clone onto filter

↓

Hybridize total mRNA with each cDNA; wash

↓

Release bound mRNA

↓

Translate mRNA *in vitro*

↓

Analyse product by immunoprecipitation or SDS-PAGE.

↓

Select clone which contained desired cDNA;
this can be bulked up and used as pure probe

**Fig. 11.8**   Hybrid-release translation

oligonucleotides of any desired sequence has come the possibility of producing synthetic probes. If the amino acid sequence of the product polypeptide is known, the various possible nucleotide sequences of its gene can be predicted. Hence, if the polypeptide is short, all possible gene sequences can be synthesized and used as probes for the natural gene. In the case of longer polypeptides this is not practicable, but fortunately a probe which is only a fraction of the full length of the gene can be used perfectly well as a probe.

The rather devious screening techniques described above are necessary because the cloned DNA remains 'dormant' in the bacterial cells, and therefore can only be detected by hybridization. However, it is possible to clone DNA in 'expression vectors' (see Section 11.2.8) in such a way that, if the DNA is a gene coding for a polypeptide product, it will be expressed to give that polypeptide, which could then be recognized by a phenotypic effect (for example, resistance to an antibiotic) or by immunoassay.

### 11.2.7  *Viral DNA and cosmid vectors*

Plasmids are excellent vectors for short fragments of DNA, and are therefore always used to build cDNA libraries, in which each DNA insert has been produced by reverse transcription of an mRNA. Since mRNA molecules are relatively short (even large proteins rarely have chains longer than about 400 amino acids, equivalent to 1200 nucleotides of mRNA), the inserted DNA fragments will not be so large that they have a serious effect on the stability or transformation efficiency of the plasmids. However, there are many instances when it is necessary to clone large DNA fragments, the most obvious being for the compilation of a genome library from a higher organism with a complex genome. It has already been pointed out that an impractically large number of colonies would have to be produced to be sure that any part of the genome was present in at least one clone, if the genome were fragmented into pieces small enough to clone in a plasmid. Thus, for the production of genome libraries, alternative vectors must be used.

11.2.7.1  *Viral DNA.* Bearing in mind that the main limitation to size of insert is the decrease in transformation efficiency, vectors for cloning large inserts must have some other mechanism for entering the host cell. This explains the importance of viral DNA, and the related 'cosmids', for such cloning. The DNA of the bacteriophage λ is 49 kb long, and is contained within a 'head' made of proteins, attached to a 'tail', also constructed of proteins. The tail is able to bind to the outer cell membrane of *E. coli*, and the linear viral DNA is then injected into the cell (Fig. 11.9). Once within the cell, the naked DNA is able to circularize, owing to the presence of complementary single strands of 12 nucleotides at each end of the molecule. These complementary ends are known as the *cos* sites, and are a key feature of the molecule. Once circularized, the molecule is replicated, and eventually, by a 'rolling circle' mechanism, gives rise to long chains of DNA composed of several complete genomes joined end to end, forming a concatamer. The λ DNA directs the synthesis of many proteins, including those which make up its viral coat. These proteins assemble in a well-defined sequence, to produce empty heads, into which the DNA is packed by a process which involves cleavage of the concatameric DNA at its *cos* sites, to yield molecules of the correct size to fit in the heads. Subsequently, tails are added and the mature viral particles are released, allowing the infection cycle to be repeated.

Fortunately there is a certain flexibility in the length of DNA which will be efficiently packaged, ranging between 79 and 109% of the length of the wild type molecule. This means that short fragments of DNA can be inserted into the λ DNA with no detriment to its replication or packaging, and so the λ DNA can be used as a cloning vector. Even more fortunate is the fact that a large number of the λ genes can be deleted without preventing growth of the virus. Consequently it is possible to excise a long stretch of the DNA and replace it with foreign DNA for cloning. Naturally the genome has had to be altered extensively to make it convenient to use for cloning; for example, restriction sites have been removed from regions of the genome which must be preserved intact.

The λ vectors which have been constructed fall into two categories: 'insertion' vectors, which contain a unique cleavage site, into which a relatively short piece of foreign DNA can be inserted, and 'replacement' vectors, which have two cleavage sites

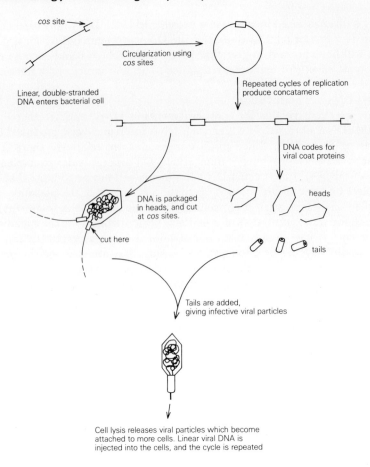

cos site

Circularization using
cos sites

Linear, double-stranded
DNA enters bacterial cell

Repeated cycles of replication
produce concatamers

DNA codes for
viral coat proteins

DNA is packaged
in heads, and cut
at cos sites.

heads

cut here

tails

Tails are added,
giving infective viral particles

Cell lysis releases viral particles which become
attached to more cells. Linear viral DNA is
injected into the cells, and the cycle is repeated

**Fig. 11.9**　Lytic cycle of bacteriophage λ

located a long way apart, on either side of a length of non-essential DNA. Cleavage at these sites will result in the formation of left and right arms, each with a terminal *cos* site, and a longer 'stuffer' region from the centre of the molecule. The unwanted stuffer fragments can be separated from the arms on the basis of their size differences, using electrophoresis or velocity gradient ultracentrifugation, and the arms can then be mixed with, and ligated to, the DNA fragments to be cloned. As will be evident from Fig. 11.10, several unwanted products could be formed by this procedure, including long chains of left and right arm pairs without an insert, or inserts made up of several short fragments linked together. The first of these products is a nuisance, since it decreases the yield of useful recombinant molecules; the second type of product could wreck attempts to build a genome library, since it would link together fragments which might be widely spaced in the original genome. By treating the insert fragments with alkaline phosphatase prior to ligation, the formation of such multiple inserts can be prevented; alternatively, by using only large DNA fragments for ligation, it can be ensured that any multiple inserts would generate a molecule too large to be packaged in the viral head.

One of the major products of the ligation should be chains of DNA made up of units of left arm/large insert/right arm, the units being linked by their *cos* sites (Fig.

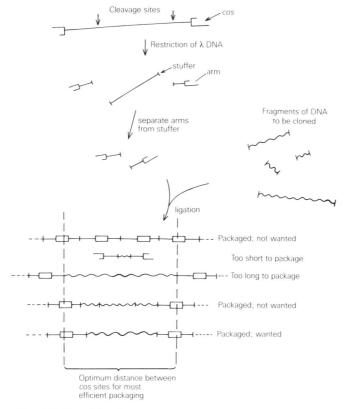

**Fig. 11.10**   Ligation products from replacement vector

11.10). Even if some of these molecules circularize, they will be too large to transform bacterial cells efficiently. However, if the molecules are mixed with a preparation of head and tail proteins, the unit lengths of DNA can be packaged *in vitro*, provided that the unit length lies within the range which can be accommodated in the head. The viral particles so formed can then be used to infect bacterial cells, into which they will inject their DNA; this is such an effective process that up to $10^8$ plaques can be obtained per μg of viral DNA. During the packaging process one 'headful' of DNA is coiled up within a newly assembled head, and this should result in the positioning of the left and right *cos* sites at the 'mouth' of the head, where they will be cleaved by a specific protein (the A protein); subsequently a tail is attached to each head. Thus the principal requirement for packaging is that a pair of *cos* sites should be present, and should be the correct distance apart. Because of this, the *in vitro* packaging provides a way of selecting for inserts of the correct length.

As with cloning in plasmid vectors, there must be ways of selecting for recombinant DNA, and for detecting particular inserts. There is now a wide choice of λ vectors, and they incorporate several features useful for selection. For example, it is possible to detect the presence of stuffer in the viral DNA if the stuffer is constructed to contain a gene coding for β-galactosidase. When *lac⁻* bacteria, which lack the β-galactosidase gene, are infected with virus, stuffer DNA will direct synthesis of the enzyme, which can be detected by its ability to hydrolyse the artificial substrate Xgal to give a blue product. Hence, blue plaques indicate the presence of stuffer, and colourless plaques will be produced when the stuffer has been replaced by an insert. Using such a test, clones containing recombinant DNA can be isolated from the non-recombinants.

To identify those clones containing a particular sequence one can use techniques very similar to those used for plasmid vectors. The colony hybridization procedure described in Section 11.2.6 can be applied to plaques as well as to colonies of cells, and so hybridization with radioactive probes can be used to 'pull out' any specific sequence of cloned DNA.

11.2.7.2   *Cosmids*. Even when the stuffer region of λ DNA has been replaced by an insert, the remaining DNA contains all the information needed for lytic growth of the virus, including the genes needed for DNA replication and synthesis of the viral proteins. In this way the DNA is cloned by repeated cycles of cell lysis and infection of surrounding cells, causing plaque formation. Since the essential genes make up about 60% of the λ genome, and the maximum length between *cos* sites which can be packaged is 52 kb, there is an upper limit on insert size of about 21 kb. Although this is a very useful length, there are occasions when even longer inserts are desirable, as, for example, in the construction of genome libraries from mammalian genomes, or for the analysis of very large genes and their flanking regions. Long stretches of DNA can be analysed by 'walking' along the chromosome by using hybridization to identify cloned sequences which overlap; obviously the rate of 'walking' will increase if larger strides are taken, and so large fragments are wanted for this technique.

To get round the size limitations of λ, a hybrid vector has been developed, which incorporates the *cos* sites of λ for packaging into viral particles, but relies on a plasmid origin of replication for DNA replication within the host cell. There are no genes for viral proteins, and so no virus particles are formed within the cells, and no cell lysis occurs. These 'cosmid' vectors therefore have the usual features of a plasmid—the origin of replication, a marker gene coding for drug resistance, unique cleavage sites for the insertion of foreign DNA, which may lie within a marker gene, and a small size. The only 'extra' is a short piece of DNA which contains the *cos* site of λ in a form where the 12 base cohesive ends are paired up and ligated. The cloning process is illustrated in Fig. 11.11, and involves linearising the cosmid with a restriction enzyme, and ligation with fragments of foreign DNA prepared using the same enzyme. As usual, a range of products will be formed, but this will include molecules consisting of foreign DNA with each of its ends joined to a cosmid. Provided that the resulting distance between *cos* sites lies in the range of 37 to 52 kb, the recombinant molecule will be packaged by an *in vitro* λ packaging system, consisting of packaging enzymes, head and tail proteins, just as if it were part of a repeating chain of λ DNA molecules. Since the cosmids have lengths of about 5 kb, DNA containing inserts between 32 and 47 kb will be packaged, and will then be transferred very efficiently into bacterial cells by infection with the virus particles. Once inside the cell, the linear DNA recircularizes by base-pairing of its cohesive ends, after which it is replicated as a plasmid. Methods of selecting for particular inserts are as described above for plasmids.

11.2.8   *Expression of cloned DNA*

Cloned DNA is of value to the molecular biologist because it provides enough material for further analysis of specific parts of a genome, for example by sequencing. However, in itself the DNA is of little use to the biotechnologist unless it can be expressed, resulting in the biosynthesis of a valuable polypeptide product.

Unless deliberate steps are taken, there is no reason why a gene cloned in a

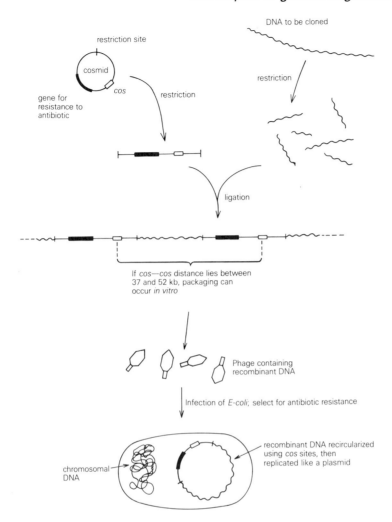

**Fig. 11.11** Cloning using cosmid vectors

plasmid vector should be expressed within the host cell. Both transcription and translation are needed to produce a polypeptide, and each process demands the presence of specific sequences in addition to those which code for the polypeptide. Transcription in prokaryotes cannot begin unless the bacterial RNA polymerase can bind to a 'promoter' region of the DNA immediately before (or 'upstream of') the gene. Two regions, 10 and 35 bp upstream of the gene are particularly important for this binding, and their precise sequence seems to determine the 'strength' of the promoter. So the presence of a suitable promoter should result in the synthesis of mRNA from a gene placed downstream of it.

In order for a ribosome to bind to the mRNA prior to translation, there must be a short sequence, about six to eight nucleotides in length, placed with its centre about eight nucleotides upstream of the initiation codon. This sequence, known as the 'Shine–Dalgarno sequence' after its discoverers, appears to be complementary to part of the 16S rRNA of the small ribosomal subunit, thus explaining its importance for ribosome binding. The gene itself must start with the universal initiation codon, AUG.

If, as is most likely, the gene is eukaryotic in origin, there may be a further problem to deal with before a functional product can be obtained. Most eukaryotic genes are made up of several sequences ('exons') which together code for the polypeptide, interspersed with other sequences ('intervening sequences' or 'introns') which are non-coding. Consequently, transcription produces an RNA molecule which is far too long, and which contains chunks of 'rubbish' in the form of introns. This RNA is processed before release from the nucleus, to yield an mRNA molecule composed entirely of the exons, which will be translated to give the correct product. Bacterial genes do not contain introns, and so it is not surprising that bacteria have no machinery for excising introns from RNA transcribed from eukaryotic genes. Consequently, correct expression of a cloned eukaryotic gene in bacteria will only be possible if the gene is one of those rare examples which has no introns. This problem disappears if the cloned DNA has been produced by reverse transcription of mRNA, as in the formation of a cDNA library, since the mRNA is already processed ready for translation.

It would be very time-consuming to add a promoter, Shine–Dalgarno sequence and (when needed) initiation codon to every gene to be expressed, especially since the relative spacing of these components is critical. This would be a particular problem in those cases where expression, followed by immunoassay of the products, is to be used to screen a large number of genes. Consequently, 'expression vectors' have been constructed, containing a strong promoter whose activity can often be controlled by temperature or the level of a specific inducer or repressor, and a Shine–Dalgarno sequence which can be at the optimum distance from an initiation codon. To aid the insertion of the gene to be expressed, there is normally at least one restriction site within, or a short distance downstream from, the initiation codon. In addition, the vectors contain an origin of replication and a marker gene coding for resistance to an antibiotic.

Several bacterial promoters have been used to construct expression vectors, including the *lac* and *trp* promoters, which are regulated by their corresponding repressors. The *lac* promoter is induced by the presence of isopropyl-$\beta$-D-thio-galactoside (IPTG), while tryptophan starvation or 3-indolylacetic acid will induce the *trp* promoter. Strength of promoter and ease of regulation make the pL promoter of $\lambda$ phage particularly attractive for use in expression vectors, since host cells can be chosen which contain a temperature-sensitive $\lambda$ repressor incorporated in their genome. Using such a system, the cells are grown to late log phase at 32 °C, after transformation with the expression vector; under these conditions the repressor is active, so the cloned gene is not expressed. Subsequent transfer of the cells to 42 °C will inactivate the repressor, and so turn on gene expression.

Provided that the inserted gene is placed in the expression vector so that its reading frame is in the correct phase with respect to the initiation codon, the correct polypeptide product should be produced. However, eukaryotic proteins are not always stable within prokaryotic cells, and in such cases it is necessary to protect the protein. This can be done by use of a vector which contains a prokaryotic structural gene placed so that it will be expressed. If a foreign gene is inserted within the prokaryotic gene, with their reading frames in phase, a hybrid polypeptide will be produced, with part of the prokaryotic product attached to the N-terminus of the foreign polypeptide. This extra sequence at the N-terminus may help to stabilize the polypeptide; however it may also interfere with the correct functioning of the foreign protein. If the product is being made for its antigenic properties, for example, to produce a vaccine, the extra amino acid residues may not do much harm, but in the majority of cases it will be necessary to separate the hybrid protein into its two parts by chemical cleavage at a specific amino acid such as methionine. Unfortunately, it is unusual to find a protein which does not contain at least one methionine within its polypeptide chain, and so chemical cleavage is inevitably limited in its applications; the method has been used for the production of 'human' insulin by bacteria. It is possible to inhibit the degradation of foreign proteins within *E. coli* by introducing the *pin* (protease inhibition) gene of phage T4 in such a way that it is expressed; in such cases there is no need to produce hybrid polypeptides.

Some proteins are 'labelled' for export from the cell by being produced as precursors with a 'signal' polypeptide at their N-terminus (for example, $\beta$-lactamase). It has been possible to direct bacteria to export foreign proteins into the medium by the creation of hybrid molecules, in which the N-terminus contained a bacterial signal sequence. In view of the relative ease of recovering proteins from culture medium, rather than by lysis of cells, the use of such signal sequences is likely to be of great importance. Similar sequences are used in eukaryotic cells to 'target' proteins for export or for deposition within specific organelles, and they are sure to find application in the genetic manipulation of higher organisms.

## 11.3  Genetic manipulation of eukaryotic cells

### 11.3.1  *Limitations of bacteria*

Genetically engineered bacteria seem to be the obvious organisms to use for the large-scale production of valuable polypeptides. They grow rapidly, are readily transformed, can use a wide range of substrates, and have relatively simple, well-characterized genomes. Consequently all the genetically engineered polypeptides so far produced on anything approaching an industrial scale have been the products of bacterial systems. However, bacteria are not without their problems. As has already been pointed out (Section 11.2.8), prokaryotes cannot remove the introns found in most eukaryotic genes, and therefore such genes cannot be expressed in bacteria. Those eukaryotic proteins that have been produced by bacteria were either coded for by genes which contained no introns, or by cDNA which had been generated from matured mRNA. Since the most valuable polypeptides are inevitably those which are normally produced in very small amounts, the isolation of their mRNA prior to cDNA synthesis is often a major obstacle in the way of obtaining bacterial production of the protein. In such cases there are obvious attractions in the use of eukaryotic cells for expression of the gene since, provided the gene is still attached to its promoter (or a suitable substitute), it should be correctly transcribed and the RNA processed to give a functional mRNA.

Another problem occurs when prokaryotes are used for the synthesis of eukaryotic proteins which require some form of alteration after they have been released from the ribosome. This 'post-translational modification' can take the form of limited proteolysis (as in the conversion of proinsulin to insulin), or the addition of oligosaccharides to specific sites on the polypeptide chain (glycosylation), either of which may be essential for the generation of an active product. Only eukaryotic cells are capable of carrying out such post-translational modifications.

Transformation of a bacterium by a plasmid is not irreversible. Although each cell usually contains many copies of the plasmid, there is no mechanism to ensure the segregation of equal numbers of plasmids into each daughter cell during division, and, particularly under conditions of rapid cell proliferation, cells will occasionally appear without any plasmids. In the absence of any selective pressure, such cells will have a slight advantage over their plasmid-containing rivals, since none of their resources will be diverted to the replication of plasmid; consequently there is a tendency for plasmid to be lost from such bacterial cultures. The usual way to ensure plasmid stability is to grow the cells in a medium containing an antibiotic to which the plasmid carries a resistance gene, thus only cells containing plasmid will survive. This is fine on a laboratory scale, but is likely to be very expensive, and even hazardous, when scaled up for fermentation on a commercial level. If the plasmid carries one or more genes which result in an increase in cell growth rate, as in the case of the glutamate dehydrogenase gene inserted into *Methylophilus methylotrophus*, grown for the production of single cell protein, then there will inevitably be selective pressure for maintenance of the plasmid. As is described below, it is now possible to construct 'mini-chromosomes' which can be used as gene vectors in yeast, and which are replicated and segregated during cell division just as if they were normal chromosomes. Certain plant and animal vectors are also maintained in a stable state within the cells by integration with the nuclear DNA.

The biotechnological use of genetic manipulation is not limited to the production of polypeptides by cell cultures. A considerable effort is going into the improvement of crops and livestock by means of genetic engineering. Inevitably this requires the development of vectors which can be used to transform eukaryotic cells and to direct the expression of the genes they carry.

### 11.3.2 *Plant cells*

11.3.2.1 *Ti plasmid.* The most promising method of transforming plant cells makes use of a plasmid called the Ti plasmid (Fig. 11.12). In nature this plasmid is found within the bacterium *Agrobacterium tumefaciens*, which lives in soil and invades many dicotyledonous plants when they are damaged at soil level. The bacterium enters the fresh wound and attaches itself to the wall of an intact cell, after which it transfers a relatively small part of its Ti plasmid into the nucleus of the plant cell. The transferred DNA is known as T-DNA, and carries several genes which are expressed within the plant and which have dramatic effects on its metabolism. One gene codes for an enzyme which catalyses the synthesis of an opine from amino acids and other common metabolites found within the plant cell. Opines are not found normally in plants, and cannot be metabolized by them; but they can be, and are, used as a substrate by *A. tumefaciens*. The particular opine produced depends on the strain of bacterium infecting the plant. Some strains, for example, result in the production of a nopaline, others cause octopine synthesis. In each case the bacterium can use only the particular opine whose production it causes. The enzymes coded for are nopaline or octopine synthases, and their respective genes are labelled *nos* and *ocs*. Not only does the T-DNA ensure a supply of nutrient for the bacterium, but it also induces the disorganized proliferation of cells around the wound to form a callus, gall, or tumour which can be further colonized by the bacteria. This disorganized growth is the result of excessive production of phytohormones, and is coded for by the *onc* gene of T-DNA.

An important property of the T-DNA is that, once inside the plant cell, it does not remain as an independent plasmid, but becomes integrated into the plant chromosomal DNA. This integration seems to depend on the presence of two repeated sequences of 25 base-pairs which are located at either end of the T-DNA, and which might join together after the T-DNA has been excised from the Ti plasmid, resulting in a circular intermediate form of the molecule (Koukolikova-Nicola *et al.*, 1985). Genes that remain on the Ti plasmid include those for attachment of the bacterium to

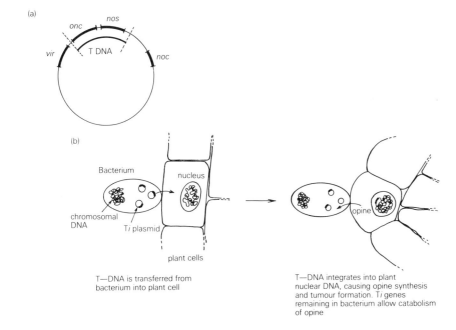

**Fig. 11.12** Structure and mode of action of T*i* plasmid. (a) Structure: important genes are labelled: *vir* is needed for transfer of T-DNA into the plant cell; *onc* causes tumour formation; *nos* codes for nopaline synthase; *noc* is needed for catabolism of nopaline by the bacterium. (b) Mode of action

plant cell walls, for transfer of T-DNA into the plant cell, and for the uptake and catabolism of the appropriate opine. The only non-integrating region of the T*i* plasmid which is essential for transfer and integration of the T-DNA is the *vir* region, which is located near to the T-DNA. Transformation of the plant cells is irreversible, and callus cells can be cultured indefinitely, long after elimination of the bacteria.

The potential of this system as a vector for the genetic manipulation of plants was quickly recognized. The genes in T-DNA are eukaryotic in nature, even though derived from a bacterial plasmid, and are transcribed by the plant's RNA polymerase; they contain introns, which are excised correctly during maturation of the mRNA. As far as the biotechnologist is concerned, little of the T-DNA is essential, since opine production and tumour formation are not required for stable integration of the DNA. Thus it should be possible to replace a *nos* or *ocs* gene with a foreign one, while retaining the opine synthase promoter to ensure expression of the foreign gene. Such experiments have been successfully carried out, resulting in the expression of genes for resistance to kanamycin and methotrexate in cultured callus cells (Schell and Van Montagu, 1983). These experiments were of great importance because they showed that T*i* plasmids could be constructed containing dominant markers for selection of transformants. It is, of course, possible to select for transformants on the basis of presence of the *onc* gene, which allows cells to grow in culture in the absence of added hormones. However, one would normally wish to regenerate an intact, healthy plant from the transformed cell, and this would be impossible in the presence of an active *onc* gene. Consequently, for regeneration of plants, the *onc* gene must be 'disarmed' by removal; and so another selectable marker gene is essential.

Since the T*i* plasmid is very large (up to 235 kilobase pairs), it is not feasible to modify it directly, so it is usual to perform all manipulations on an excised piece of DNA including the T-DNA, and then use *in vivo* recombination to swap the 'engineered' T-DNA for its normal version in an intact T*i* plasmid. Some simpler vectors have already been developed, based on the knowledge that only the T-DNA and *vir* regions are essential for transformation, and there is sure to be much progress in this respect.

Although it is convenient to use opine promoters for the expression of foreign genes (assuming the gene is inserted in the correct reading frame), they do not allow the selective expression of such genes, since they are permanently active. In practice it is likely that 'engineered' genes, like most normal genes, would need to be expressed only at certain stages of development or in specific tissues. This will require the use of specific, regulated promoters, and there is evidence that the opine synthase promoters can be successfully replaced by controllable promoters of plant origin. The gene for bean phaseolin has been expressed in sunflower cells after transformation with T*i* plasmid in which the phaseolin gene was under the control of its own promoter (Murai *et al.*, 1983). Since phaseolin is synthesized specifically in developing seeds, one would not expect the phaseolin promoter to be active in undifferentiated sunflower cells, and it was indeed found that the level of transcription of the gene was much lower when using its own promoter than when the gene was attached to the *ocs* promoter. Another important issue arising from this work was the discovery that the phaseolin produced in sunflower cells was rapidly degraded to polypeptide fragments, presumably because it could not be packaged within protective protein bodies, as would occur in developing bean seeds. Clearly the genetic engineer's problems do not always end with the successful translation of a desired mRNA.

*A. tumefaciens* does not induce tumours in monocotyledonous plants, and since most of the major crop plants, including cereals, are monocots, the commercial potential of the T*i* plasmid might seem to be rather limited. However, a glimmer of hope has been provided by the discovery that *A. tumefaciens* can transfer its T-DNA into certain monocots, resulting in expression of the opine gene within the plant cells, but without inducing tumour formation (Hooykaas-Van Slogteren, Hooykas and Schilperoort, 1984). If the T-DNA becomes integrated into the plant chromosomal DNA, and if similar results can be obtained using cereals, then the T*i* plasmid will, ironically, be even more suited to the transformation of monocots than dicots, since there seems to be no need to disarm the *onc* gene when infecting monocots.

11.3.2.2  *Cauliflower mosaic virus.* Another potential plant vector is the DNA of cauliflower mosaic virus (CaMV). This has several features which are in complete contrast to those of T*i* plasmid, some of which make it quite attractive as a vector. One useful feature is that the naked DNA is infective, being able to enter plant cells directly if rubbed onto a leaf with a mild abrasive. Once inside the cells, the DNA is replicated and encapsidated within virus particles, which then invade the rest of the plant. Although the CaMV DNA does not become integrated into the chromosomal DNA, and is therefore not certain to be handed on to all cells during cell division, its spread throughout the plant means that transformed plants can be effectively cloned by vegetative propagation.

Unfortunately, these advantageous characteristics are offset by the fact that, probably because of the need for encapsidation, the size of the CaMV genome cannot be increased significantly; yet almost all of the genome is essential, and cannot be deleted to make room for inserted foreign DNA. This restriction in capacity, coupled with the very limited host range of the virus, makes it unlikely that CaMV will be a particularly useful vector. However, some of its components, such as its very strong promoters, may be of value when incorporated in other vectors.

11.3.2.3 *Direct transformation*. A recent development, which those groups who have spent years working on plant vectors will no doubt be watching with interest, is in the use of 'direct' transformation of plant cells by DNA fragments (Paszkowski *et al.*, 1984). It has been shown that plant protoplasts treated with polyethylene glycol, which is commonly used to induce protoplast fusion, will take up DNA from their surrounding medium. More importantly, this DNA can then be stably integrated into the plant chromosomal DNA. Using this technique, it has been possible to transform plant cells with a gene for kanamycin resistance linked to a strong plant promoter. The gene was expressed, and so transformed cells could be selected using a culture medium containing kanamycin. Plants regenerated from the transformed cells were resistant to kanamycin, and this resistance was heritable. There appears to be no favoured site for the integration of transforming DNA, and randomly varying lengths of DNA appear to be integrated. However, as long as there is a way of selecting for those transformants which have the whole of a desired gene integrated, this should not present a great problem.

### 11.3.3 *Mammalian cells*

The large-scale culture of mammalian cells is far more costly and difficult than that of bacterial, yeast, or even plant cells, and their commercial use is therefore limited to the preparation of molecules which are natural products of the animal cells, such as viral particles (for vaccines), hormones or monoclonal antibodies. The value of such products can justify their high cost of production, but it seems likely that yeast and bacterial cultures will increasingly take over from animal cells.

11.3.3.1 *Direct transformation*. The most straightforward way of getting foreign DNA into a mammalian cell is to precipitate the DNA with $Ca^{2+}$, and mix this precipitate with the cells to be transformed. The DNA is taken up by the cells and, once inside, the transforming fragments are ligated to give a concatamer, which is then integrated as one large block into the nuclear DNA. As with the direct transformation of plant cells, integration occurs at random. The formation of concatamers is a particularly useful feature, since a selectable marker gene can be mixed with a gene to be cloned, in the knowledge that they will be linked together prior to integration. Several marker genes have been developed which allow positive selection of transformed cells; one of these allows cells to use xanthine as a precursor for the synthesis of purine nucleotides when the normal pathway from hypoxanthine has been blocked by the presence of mycophenolic acid in the medium. Another marker gene endows the cells with resistance to neomycin or an analogue called G418, which would normally inhibit protein synthesis.

11.3.3.2 *Viruses*. The efficiency of transformation of animal cells can be greatly increased by using viral DNA as a vector, and allowing the viral particles to insert their DNA into cells they infect. A further attraction of viruses is that they contain very strong promoters, which can be used to ensure the expression of inserted, foreign genes. The most widely used animal virus is SV40, which contains a circular DNA molecule about 5.2 kb in length. This DNA, in addition to a replication origin, consists of 'early' genes, which are needed for replication of the DNA, and 'late' genes, which code for viral coat proteins. Foreign DNA can replace either the early or late genes, provided that the host cells are coinfected with a 'helper' virus, containing functional copies of the missing genes. If a cell line known as *COS* is used as host, there is no need to use a helper virus to supply missing early genes, since these genes have been integrated into the nuclear DNA. This overcomes the problem of separating recombinant and helper viruses from each other after their recovery from the cells.

11.3.3.3 *Microinjection.* The transformation of cloned animal cells cannot be used for breeding purposes, since it is impossible to regenerate whole animals from such cells. Instead, DNA must be transferred into the nuclei of fertilized eggs, using micro-injection. It is found that injected DNA will integrate at random into the nuclear DNA, and, if an injected gene is attached to a suitable promoter it *might* be expressed. However, it is clear that the position of inserted DNA within the chromosomes has a profound influence on whether it is expressed, making the results of such manipulations highly unpredictable. After microinjection, the egg must be reimplanted in a surrogate mother, and only after gestation can the progeny be screened for expression and correct regulation of the foreign DNA. This is therefore a slow, labour-intensive method of genetic manipulation, and the small size of population which can be produced for screening inevitably reduces the chances of success. Nevertheless, this method has been used to transfer the gene for rat growth hormone into mice, in a few of which the gene was expressed, resulting in the production of 'giant' mice.

### 11.3.4 *Yeast*

Because of their slow growth rates neither plant nor animal cell cultures are suitable for the large-scale biosynthesis of polypeptides, and so bacterial cultures are generally used for such purposes. However, bacteria have their limitations. As has already been discussed, eukaryotic genes frequently contain introns, so, unless the corresponding mRNA can be isolated for synthesis of cDNA, such genes will not be expressed in prokaryotic cells. Even if expression is obtained there may be a need for post-translational modification of the polypeptide in order to generate an active product; this will not occur in bacterial cells. Bacteria also have the disadvantage that, even if they are not themselves pathogenic, they may be able to exchange some of their DNA with cells which are pathogens, either directly or through a transmissible virus, and so they must be treated as potential pathogens, even though the use of disabled host strains and plasmids affords a high level of protection.

Some of these limitations to the application of bacteria can be overcome by the use of yeast (*Saccharomyces cerevisiae*), which is a eukaryote with a small, well-characterized genome. Yeast has a very much faster growth rate than animal and plant cells, is non-pathogenic, and is not known to exchange DNA with any pathogen. Many of its genes contain introns, which are spliced out during processing of the yeast mRNA. However, it appears that these introns contain sequences necessary for correct splicing, which are not found in the introns of higher eukaryotes; and so it is still necessary to use cDNA when aiming for the expression of many foreign genes in yeast. A major advantage of yeast over bacteria is that it will carry out post-translational modifications, such as the removal of a signal sequence from a precursor polypeptide as it is secreted from the cell. It will also glycosylate polypeptides, although the nature of the glycosylation may not be identical to that produced by animal cells. As discussed below, the stability of cloned DNA can be enhanced by the formation of mini-chromosomes, and the possibility of direct integration of DNA makes yeast particularly useful for site-directed mutagenesis (see Section 11.4). As more is found out about the genetic manipulation of yeast, it becomes increasingly attractive as a host for the expression of eukaryotic genes.

There are several approaches to cloning in yeast, all involving the uptake of foreign DNA by the cells. This can be achieved by digesting the cell walls enzymatically to produce spheroplasts, exposing the spheroplasts to DNA in the presence of $Ca^{2+}$ and polyethylene glycol, and allowing regeneration of cell walls after the DNA has been taken up. If the foreign DNA is to be maintained within the growing and dividing yeast cells it must be attached to DNA containing a sequence which is recognized by the yeast as an origin of replication. Hence bacterial plasmids would not be expected to undergo replication in yeast, since their origins of replication are prokaryotic in nature. However, yeast contains its own plasmid, known as the $2\mu$ circle, which is maintained at levels of about 50 copies per cell. In principle this could be used as a vector for cloning foreign genes, but in practice it has proved easier to isolate its replication origin and the REP genes which prevent the plasmid copy number from falling too far, and attach these to the DNA to be cloned. Such plasmids usually also contain a gene which complements a defective gene in the strain of yeast being used as host, thus allowing selection for transformed cells.

It has been found that certain sequences of DNA will be replicated in yeast very efficiently, along with any DNA to which they are attached. Such sequences are known as 'autonomously replicating sequences' or ARS fragments, and it is not known if they are true origins of replication in the molecules from which they are derived, or if they fortuitously act as origins in yeast. Plasmids can be constructed using an ARS to ensure replication in yeast, and a selectable marker to ensure maintenance of the plasmid in yeast cultures grown under selective conditions. By incorporating a length of DNA which contains a yeast centromere (the CEN region) in the plasmid, segregation can be more controlled, since the plasmid is then effectively a small chromosome and becomes attached to the spindle apparatus during cell division. Unfortunately, chromosomes become less stable the smaller they are, and so such 'mini-chromosomes' may still be lost from cells eventually.

By including a bacterial origin of replication in a yeast plasmid a 'shuttle vector' is created (Fig. 11.13), which can be conveniently manipulated and cloned in bacteria, and then transferred to yeast for possible expression of inserted eukaryotic genes.

There is no need for any type of replication origin if the foreign DNA can become integrated into the yeast chromosomal DNA, since the integrated DNA will be replicated as part of a chromosome. Such integration occurs by a specific crossing-over between homologous regions of the chromosomal and integrating DNAs, as will

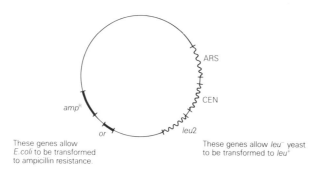

These genes allow
*E.coli* to be transformed
to ampicillin resistance.

These genes allow *leu⁻* yeast
to be transformed to *leu⁺*

**Fig. 11.13** Yeast/*E. coli* shuttle vector. *Key: amp*$^R$: ampicillin resistance; *or*: prokaryotic origin of replication; ARS: autonomously replicating sequence; CEN: yeast centromere; *leu*2: complements a defective yeast gene, allowing growth in the absence of leucine

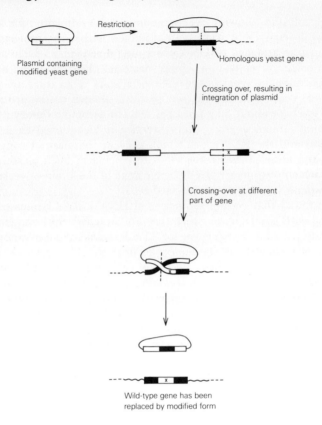

**Fig. 11.14**   Gene replacement in yeast

occur, for example, if a plasmid contains a copy of a yeast chromosomal gene (Fig. 11.14). The requirement for homology means that by attachment of an appropriate sequence to the foreign DNA, it can be precisely directed to a particular site on the chromosomal DNA; this is in contrast to the insertion of DNA into the chromosomes of animal and plant cells, which seems to occur at random. Since the insertion is reversible, and either the chromosomal or 'foreign' copy of the homologous region may be evicted, this method can be used to replace a wild-type gene with a modified version, thus opening up the prospect of 'enzyme engineering' in yeast. There is no need for the inserted DNA to be in the form of a circular plasmid; in fact integration is greatly enhanced if the homologous regions are at either end of a linear molecule, as would be produced by cutting a plasmid within the homologous sequence.

The genetic manipulation of eukaryotic cells is sure to increase in importance, both for the 'improvement' of higher organisms and for the use of cultured cells as factories for the biosynthesis of valuable products.

## 11.4   Site-directed mutagenesis

Techniques which allow the introduction of mutations at specific places in a genome have been of enormous value to molecular biologists, helping in particular to identify regions concerned with the control of gene expression. A common approach has been to clone a substantial length of DNA, including the region of interest, and then to manipulate the cloned DNA *in vitro* in such a way that mutant forms are produced. Short lengths of DNA can be deleted, or insertions or substitutions made in the DNA; these changes are often made at random, generating a mixture of mutant DNA molecules. The DNA is then reintroduced into cells, which are screened for alterations in the level or nature of the particular gene product. The DNA of altered cells can then be examined to locate the mutation, and hence assign functions to each region of the gene.

There is a wide choice of methods for introducing short deletions or insertions into a piece of cloned DNA. In both cases, the first step is to cut the DNA in at least one place, either randomly, or, if there is a conveniently placed restriction site, at a specific position. Deletions can then be obtained by limited enzymic degradation of free ends of the DNA, yielding blunt ends which can subsequently be religated. This regenerates the original molecule, minus a short region on either side of the cleavage site. Insertion can be achieved by adding linkers to the cut ends of the DNA, followed by restriction of the linkers to generate cohesive ends, and finally ligation to regenerate molecules with a few extra nucleotides derived from the linkers inserted at the cleavage site.

Even when insertions or deletions are directed to a specific site on the DNA, their effects are likely to be fairly radical, and therefore such mutations are unlikely to be of use in making slight modifications to the sequence of a gene. However, it is precisely such subtle modifications that are likely to be of greatest use to biotechnologists, who may wish to alter a single amino acid in a protein in order to tailor its properties to their needs (see Sections 14.1.3 and 14.5.2). Such alterations require the introduction of point mutations, in which a single nucleotide is changed, at a unique position in the gene. Most of the methods for creating point mutations by substitution of one nucleotide for another can only be used if the mutation is very close to a restriction site. For example, in the presence of ethidium bromide, which binds to DNA, restriction enzymes will nick the DNA by cutting only one strand at their restriction sites, and these nicks can then be opened up to produce single-stranded gaps about five nucleotides in length. By use of nucleotide analogues (which are relatively non-specific in their base-pairing), or by omitting one of the four deoxynucleoside triphosphates, the gaps can be filled in again in such a way that mismatching occurs in one position within the gap. After replication of the altered DNA, approximately half the resulting molecules will be wild-type, and half mutant.

Probably the most powerful method of introducing point mutations into a gene is known as 'oligonucleotide-directed mutagenesis' (Fig. 11.15). This technique can only be used when the nucleotide sequence of the gene is known, but, in practice, such information will be available for any gene whose product is sufficiently well characterized for its modification to be attempted. When it has been decided which base is to be changed, an oligonucleotide is synthesized corresponding to the mutated nucleotide and its neighbouring regions, typically between 15 and 20 nucleotides in length. This oligonucleotide is allowed to hybridize with a single stranded clone of the wild-type gene, produced using an M13 cloning system. Provided this hybridization is performed under conditions of 'low stringency' (that is, at low temperature, in a high concentration of salt, resulting in some tolerance to mismatched base-pairs), the oligomer will base-pair with its wild-type complementary sequence, and can then be used as a primer for the production of double-stranded DNA by DNA polymerase. One strand of this DNA will be wild-type, and the other will contain the desired point mutation. Replication of the DNA in a bacterial host will result in a mixture of

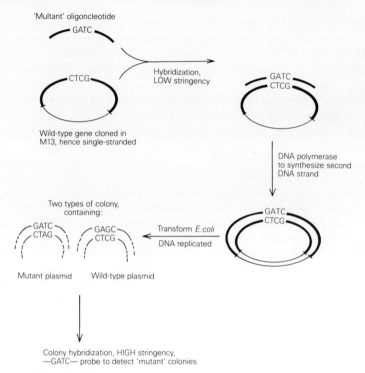

**Fig. 11.15**   Oligonucleotide-directed mutagenesis

wild-type and mutant double-stranded DNAs, which can be extracted and used to transform more cells, generating clones which contain either wild-type or mutant DNA. Fortunately it is quite easy to distinguish between the two types of clone: DNA from lysed colonies is immobilized on nitrocellulose filters and hybridized with radioactively labelled oligonucleotide under conditions of 'high stringency' (high temperature and low salt concentration), in which only mutant DNA will hybridize with the probe. Thus colonies containing the mutated DNA can be identified and isolated for further culture.

The potential of oligonucleotide-directed mutagenesis for 'protein engineering' has been clearly demonstrated using tyrosyl-tRNA synthetase (TyrTS) as a model enzyme. A group of enzymologists and X-ray crystallographers based at Imperial College, London, was able to characterize the enzyme in enough detail to allow prediction of amino acid substitutions that might increase the affinity of the enzyme for its substrate, ATP. A threonine side-chain was identified as playing a role in the binding of ATP, and it was suggested that, if this could be replaced by alanine or proline, the enzyme's affinity for ATP might increase. On the basis of these predictions, molecular biologists at Cambridge used oligonucleotide-directed muta-genesis to construct mutant TyrTS genes in which a specific threonine codon was replaced by one coding for either alanine or proline. When these mutant genes were expressed, and the resulting TyrTS enzymes analysed, it was found that the mutation to alanine resulted in a small increase in affinity for ATP, but the substitution of proline caused a dramatic increase in affinity. Thus, at low concentrations of ATP, the 'engineered' enzyme would be far more active than the wild-type (Wilkinson *et al.*, 1984). It is tempting to think that this is an example of man improving on nature, but, as the 'engineers' themselves point out, the normal intracellular levels of ATP are so high that both mutant and wild-type enzymes are working close to saturation with

ATP, so affinity is not a rate-limiting factor *in vivo*. In fact, the mutant forms have reduced catalytic rate constants for the aminoacylation reaction, and so would be less active than the wild-type when ATP levels are nearly saturating. We can expect enzymes to be perfectly suited to their roles within the cell, but there is no reason why they should have evolved to suit the conditions used for industrial processes. The importance of the TyrTS work is that it allows us to be very optimistic about the prospect of 'tuning' existing enzymes, in terms of affinity, catalytic rate constant, temperature optimum, pH optimum, etc., to match the requirements of the biochemical engineer.

## 11.5  Summary

The process of gene cloning consists of the following major steps: isolation of a gene, insertion of the gene into a vector, transfer of recombinant vector into a bacterial cell, selection of cells containing the desired recombinant vector, growth of selected cells, and use of those cells (if the gene is expressed) or recovery of the cloned gene from them.

Each of the above steps is considered in detail. Restriction enzymes are used to cut DNA at specific sites. Messenger RNA can be isolated and used to direct the synthesis of cDNA, which can then be cloned to produce a gene library. Plasmid cloning vectors contain genes for antibiotic resistance, and these usually contain sites which can be cut by a restriction enzyme prior to insertion of a piece of DNA to be cloned. Ligation is necessary to form a covalent link between vector and insert. The resulting recombinant DNA is taken up by bacterial cells, which are said to be transformed. After growth of transformed cells, those colonies which contain a particular recombinant DNA are selected, using colony hybridization. This process demands the preparation of a 'probe' complementary to the desired DNA. Complementary DNA (cDNA) is frequently used as a probe, and can be isolated from a cDNA library using hybrid-release or hybrid-arrested translations. Expression vectors can also be used for detecting a specific cDNA.

Viral DNA and cosmid vectors are used instead of plasmids for cloning large inserts, as, for example, when preparing a genome library. The DNA is packaged within viral head particles by a process which involves the *cos* site, and infection of bacteria is followed by very efficient transfer of DNA into the bacterial cells.

Expression of cloned genes can be achieved, but, if the gene is eukaryotic in origin, it may be necessary to remove introns. Expression vectors have been constructed, and are discussed. Export of proteins from the cell can be of great importance in facilitating their recovery.

The genetic manipulation of eukaryotic cells has some advantages over the use of bacteria, especially when post-transitional modification is important, and it can be used for crop and livestock improvement. For the genetic manipulation of plants T*i* plasmid can often be used as a vector. DNA can also be introduced into plant cells using cauliflower mosaic virus, or by the direct transformation of protoplasts. Mammalian cells are manipulated using direct transformation, viral DNA vectors, or microinjection. The genetic manipulation of yeast is of great importance since it is fast-growing, safe, cheap, and eukaryotic. Transformation of spheroplasts by yeast vectors may be followed by integration at homologous sites in the chromosomal DNA.

Site-directed mutagenesis, in which point mutations are introduced into a specific position in a gene, allows the biotechnologist to undertake protein engineering, with the aim of altering the properties of enzymes to suit industrial requirements.

# Chapter 12

## *Achievements of and prospects for genetic engineering*

### 12.1 Achievements

What are the achievements of genetic engineering in the area of biotechnology? So far very few genetically engineered products have gone into production commercially, although many have reached the pilot plant stage, or are in clinical trials. Genetically engineered bacteria have been used for the production of single cell protein for several years. 'Human' insulin made by bacteria is available, but faces tough competition from chemically modified pig insulin. The only vaccines so far available commercially are for use on livestock, where there is a large demand. Such vaccines cannot offer the prolonged protection which is given by attenuated, live bacteria, but they are safer to use. These few products, however successful, cannot possibly generate enough profit to cover the vast amount of money which has already been invested in the development of genetic engineering, and it is certain that investors will have to wait for a few more years before seeing a worthwhile profit. Already a few firms have found themselves in financial trouble because of a lack of appreciation of the time needed to overcome the formidable problems associated with the development, production and testing of a new product. The United States and Japan have shown by far the greatest commitment to genetic manipulation, and they can be expected to start reaping the very considerable rewards of their foresight within the next few years. In fact it is expected that $40 billion of sales worldwide will be attributable to genetic engineering by the year 2000 (Gregory, 1984), so there is every incentive for investing in research now.

The projects which are likely to have both commercial and social benefits in the near future include the production of human blood clotting factor VIII, interferons, tumour necrosis factor, lymphotoxin, and vaccines against herpes simplex, hepatitis B and other human viruses. Bacteria will be available to degrade pollutants such as 2,4,5-T or DDT, or to produce surfactants, emulsifiers and other specialized molecules. All these projects have been successful on a laboratory scale, and must now be scaled up.

## 12.2  Problems

### 12.2.1  *Expression and plasmid stability*

Genetic engineering has reached the stage where it is relatively straightforward to isolate and clone the gene for any protein which has been well characterized. The time-consuming problems arise after the cloning stage. To begin with there is the need to obtain high levels of expression of the gene in a foreign host, ideally under some form of control, so that expression can be delayed until the host cells have been able to grow to a high cell density. The protein product of the gene might not be stable within the host cell, so it may be necessary to use a hybrid gene as a way of 'tagging' the protein with a protective polypeptide; alternatively, the genome of the host cell could be modified to prevent degradation of foreign proteins. If the foreign gene is on a plasmid, steps must be taken to ensure that the plasmid is maintained in the cell culture. When the foreign gene endows the host cell with an increased efficiency of growth, as in the case of the glutamate dehydrogenase gene transferred to *M. methylotrophus*, plasmid stability is automatically ensured; but, in the majority of cases, another gene must be included in the plasmid which will apply a selective pressure in favour of plasmid-containing cells.

### 12.2.2  *Product recovery*

Even when expression of a gene on a stable plasmid has been achieved our worries are not over. Ease of recovery of a product may be a major factor in determining whether a process can be profitable. If the product remains within the bacterial cells, the cells must be harvested, broken open, and the debris separated from product. On the other hand, if the cells can be persuaded to secrete the product, its recovery becomes much simpler, with the bonus that cells are not destroyed during the process. Eukaryotic cells will secrete protein with a suitable signal sequence at its N-terminus, removing the signal sequence as the protein leaves the cell, but prokaryotes are generally less able to secrete proteins. However, *Bacillus subtilis* is known to secrete more than 50 proteins, and it is therefore often used in preference to *E. coli* when secretion is important. The problems of recovering enzymes from cultured cells are discussed in detail in Chapter 13.

### 12.2.3  *Safety*

Another factor which must be considered carefully is the safety of any genetically engineered system. When cloning began in the mid-1970s there seemed to be a real possibility that harmful genes (such as those coding for tumour formation, or for a toxin) could be inadvertently transferred to bacteria which would survive outside the laboratory, and which might then spread amongst the population. Strict guidelines were drawn up by the National Institutes of Health in the United States and by the Genetic Manipulation Advisory Group in the United Kingdom, effectively banning the cloning of viral DNA or DNA from tumours, and specifying high levels of physical containment for most other cloning experiments. Fortunately, 'safe' host cells were developed, such as 1776, which had such specific nutrient requirements, weak cell walls, high sensitivity to bile salts, etc., that they were highly unlikely to survive outside the laboratory. At the same time, plasmids were constructed which lacked the *tra* genes necessary for transfer of the plasmid from one cell to another by conjugation, thus reducing the danger of transfer of a dangerous plasmid from a safe host to an infective one.

In order for a cloned gene to become harmful, *all* the following must occur. The gene must be transferred to an infective host. The new host must escape from containment and cause infection. The cloned gene must be expressed by the host, and the product must be able to move from the bacterial cell into the infected person, where it might cause damage if it reaches a particular location. Since the overall probability of such an incident is the *product* of the probabilities of each step, the risks start to look fairly insignificant, provided that disabled hosts and plasmids are used, with reasonably strict levels of physical containment. In recognition of this, the guidelines have been progressively relaxed, and the only serious restrictions which remain relate to the release of genetically engineered organisms into the environment, as might be needed for the microbiological degradation of pollutants, or for microbially enhanced oil recovery. Work with pathogens must still, of course, be carried out under strictly regulated conditions of containment.

### 12.2.4   *Economics*

There is no guarantee that a product will be profitable just because it is made by genetically engineered organisms. If the market is very limited there is no possibility of making a profit, unless the customers are all oil sheiks. Such an argument could apply to human growth hormone, since there are relatively few cases of dwarfism worldwide, and therefore insufficient potential revenue to cover development and production costs. However, if it is found that the hormone is also effective in treating burns and in promoting healing, then the market will be vast, and production of the hormone should be most profitable, particularly since demand would then far outstrip the supplies available from cadavers. This single example should make it clear that much of the investment in genetic engineering must be speculative, except where there is an obvious demand for, and no satisfactory alternative source of, a product. Even in such cases there can be some horrible pitfalls, among which is the problem of adequate patent protection for any new recombinant DNA. Although it is possible to patent an organism or a plasmid, it is not yet clear how much protection is given against the creation of cells or plasmids which do the same job but in a subtly different way; for example, some of the bases in a gene can be altered without changing the protein for which it codes.

### 12.2.5   *The need for basic research*

One of the main obstacles in the way of the genetic engineer has nothing to do with gene manipulation; it is the fact that we are still depressingly ignorant about many of the basic principles underlying the features we would like to modify. Until we have a thorough understanding of the biochemistry determining disease resistance, growth rates, morphology, photosynthetic yield, etc., it will be difficult to manipulate such characteristics by genetic engineering.

## 12.3   The future

In this book, genetic manipulation has been discussed only in relation to its direct application to biotechnology; however, it should not be forgotten that a principal application of these techniques is to the investigation of gene structure and function. Such basic studies will inevitably have a great influence on biotechnology, with respect both to what is cloned and how it is cloned. We can expect to learn a lot more about the factors involved in regulating gene expression, and should be able to make use of this knowledge in the optimization of expression of foreign genes. A full understanding of the mechanisms of secretion and post-translational modifications will help in the efficient production of fully active and stable polypeptides.

What developments can we hope to see over the next ten years? Life would be very dull without surprises, and there is no reason to think that genetic engineering will fail to present us with some. However, on the basis of the work currently being carried out, it is possible to make some guesses about which general areas are likely to produce results.

### 12.3.1 *Pharmacology*
Cloning will produce enough material to allow full clinical trials of those polypeptides, such as interferons, enkephalins, cytokines, etc., which are naturally produced at a low level within the body, and which probably act as 'biological response modifiers'. If these could be used as 'natural' drugs, a whole new area of pharmacology could be opened up. The number of vaccines produced by genetic engineering seems certain to increase, and should include several for use in humans; there are hopes for vaccines against malaria and AIDS.

### 12.3.2 *Industrial enzymes*
As the industrial use of enzymes becomes more important (see Chapter 14) we can expect that they will be produced by genetically engineered micro-organisms. Each gene will probably be altered by *in vitro* mutagenesis to give an enzyme which is modified to improve its stability, its kinetics, and perhaps to suit it to being immobilized for most efficient use on an industrial scale. Such 'protein engineering' is likely to be of immense importance to the development of enzyme technology, and is entirely dependent on the techniques of genetic engineering. One project which is already making significant progress is the production of ligninases, which could make available a vast new source of chemical feedstock, based on the degradation of woody waste. For further discussion of the potential uses of enzymes produced by genetic engineering see Chapter 15.

### 12.3.3 *Breeding*
At the level of whole organisms, we can look forward to some significant improvements in the quality of crops, in terms of decreased need for fertilizer and pesticides, improved disease resistance, increased yield, and improved quality of plant proteins. Similar improvements in the quality of livestock may be possible. There are already reports of the synthesis of novel compounds by bacteria, as a result of the creation of new metabolic pathways. For example, novel 'hybrid' antibiotics have been synthesized by *Streptomyces* as a result of bringing together in one organism the biosynthetic genes from strains which produce different antibiotics (Hopwood *et al.*, 1985).

### 12.3.4 *Alternatives*
It has been argued (Vane and Cuatrecasas, 1984) that the production of polypeptides by genetic engineering will be of only transient importance, since advances in our understanding of receptor–effector interactions may allow the design of small molecules which will be as effective as large polypeptides, yet which can be synthesized chemically. It is true that short polypeptides between about 8 and 20 residues in length have been found to be more effective than whole viral proteins in eliciting the production of antibodies to foot and mouth disease virus or poliovirus, and many bioactive polypeptides are very short (for example, calcitonin 32 residues, endorphins

31, enkephalins 5). Chemical synthesis offers the possibility of modifying some residues, for example by using D-amino acids, so that the polypeptide will not be degraded within the gut, thus allowing oral administration of drugs which would otherwise have to be taken intravenously. The choice between biological and chemical synthesis will be influenced by such factors as technological advances in large-scale polypeptide synthesis, the costs of raw materials, the degree of purity required, etc., and it is impossible to predict how long it will be before drugs can be rationally designed to suit the receptors to which they bind. It is likely that genetic engineering will be used for the production of polypeptides for many years to come, and for enzyme production it will remain the only feasible method for the foreseeable future. Genetic engineering is already becoming a tool of the breeders, and its importance in this area is certain to increase.

## 12.4   Summary

A survey of achievements includes the biosynthesis of insulin, vaccines, blood clotting factors, antiviral and anticancer factors.

The main problems which remain to be solved are: efficient and controlled expression of cloned genes; plasmid stability; recovery of product.

The new technology presents some potential risks. Consequently, guidelines have been drawn up, and 'safe' host cells and vectors have been developed.

Economics will inevitably determine the future of genetic engineering, and there must be sufficient demand for a product to justify the high investment needed to bring it to the market. It is not yet clear how effectively a genetically engineered product can be protected by patents.

A brief survey of potential products from genetic engineering includes 'biological response modifiers', vaccines, enzymes for industrial use, improved crops and livestock, production of novel compounds by bacteria. However, it has been suggested that the chemical synthesis of oligonucleotides may challenge their production by genetic engineering.

# SECTION V

# Enzyme Technology

*M. D. Trevan*

## Assumed concepts

Basic principles of protein structure
Michaelis-Menten enzyme kinetics
Effect of pH and temperature on enzyme-catalysed reactions
Competitive and non-competitive inhibition
Chromatographic separations
Electrophoretic separations

*Useful books*
Godfrey T. & Reichert J. (1983) *Industrial Enzymology*. London, MacMillan.
Mosbach K. ed. (1976) *Methods in Enzymology, Vol XLIV*. New York, Academic Press.
Palmer T. (1981) *Understanding Enzymes*. Chichester, Ellis Horwood.
Stryer L. (1981) *Biochemistry*. San Francisco, Freeman.
Trevan M. D. (1980) *Immobilized Enzymes*. Chichester, John Wiley.

# Chapter 13

## *Enzyme production*

### 13.1  Introduction: The use of enzymes

The practical application of enzyme catalysis is big business. The total world market for enzymes in 1981 was estimated at 65 000 tonnes with a value of $400 × 10^6$; by 1985 this was expected to grow to 75 000 tonnes ($600 × 10^6$). Enzymes are used in four distinct fields as: therapeutic agents; manipulative tools, for example in gene manipulation; analytical reagents; and 'industrial' catalysts. The largest of these market areas is, and will remain for the forseeable future, the use of enzymes as industrial catalysts. Thus while there are some 25 companies engaged in enzyme production in the western world, only a small handful are involved in the bulk of the production. Novo Industries of Denmark produces 50% of the enzymes sold on the 'western' market place, Gist-Brocades of Holland a further 20%, while the total output of the United States accounts for only 12% of the total. This is not to imply however that this distribution of volume production necessarily reflects the value of the market, for the high volume producers tend to produce the low value enzymes for which there is an existing high demand, principally for use as industrial catalysts. Conversely the smaller companies specialize in the production of low volume, high value enzymes principally used at present by those involved in the use of enzymes as manipulative or analytical tools. As an example of this one of the cheapest enzymes available commercially, a bacterial $\alpha$-amylase, costs just $1.7 kg$^{-1}$ and has a market of 3000 tonnes per annum. At the other extreme, ornithine carbamyl transferase costs $215 × 10^6$ kg$^{-1}$ and probably sells less than 10 mg per annum.

It is no surprise, therefore, that 80% of the enzymes produced annually are simple hydrolytic enzymes and of these 60% are proteases. Table 13.1 shows the approximate percentage volume production for different classes of enzymes. From this it can be observed that over three times the quantity of enzymes ends up in either cheese or washing powder as is used in total in the analytical, pharmaceutical and developmental fields put together! It would be appropriate to conclude therefore that washing and eating (i.e. the production of detergents and food proceesing) are the 'bread and butter' of enzyme technology at present.

However, when considering the future potential of enzyme technology it is likely

**Table 13.1** Distribution of industrial enzymes

| *Proteases* 59% | | *Carbohydrases* 28% | | *Lipase* 3% | *Other* 10% |
|---|---|---|---|---|---|
| Alkaline (detergents) | 25% | α-Amylases | 13% | | Analytical |
| Neutral | 12% | Isomerases | 6% | | Pharmaceutical |
| Acid (rennets) | 10% | β-Amylases | 5% | | Developmental |
| Alkaline (other) | 6% | Pectinases | 3% | | |
| Trypsins | 3% | Cellulases | 1% | | |
| Acid (other) | 3% | Lactase | | | |

that the major growth will be in the non-traditional uses of enzymes: as analytical reagents in fully automated analysers for use in clinical chemistry, effluent monitoring and process control; as industrial catalysts to produce fine chemicals at present produced by purely chemical means. It is in these areas therefore that this section will tend to concentrate.

One final caution must be issued by way of introduction and that is to warn the reader what may be implied by the term 'enzyme' in this context. If one set out to buy some commercially produced enzyme, one might at one extreme end up with a product which contains only 10% protein and only 1 or 2% active enzyme which itself might be a mixture of similar enzymes. For example the alkaline proteases used in washing powders are commonly supplied as 2–3% active enzymes 'packaged' with 10% other protein and 85% inorganic salts, the principal components of which would be bicarbonate and phosphate. As we shall see there is good reason for this. At the other extreme the enzyme will be virtually 100% of a pure single molecular species (for example, endonucleases used in gene manipulation). In addition the use of whole possibly modified, microbial or plant cells to catalyse specific conversions is becoming increasingly common. Thus there is a continuum of type of 'enzyme' in use as industrial catalysts from the high purity enzyme, through the modified cell catalyst eventually merging into the area of fermentation technology. The micro-organism used in a classical fermentation is in essence a complex multienzyme catalyst. For these reasons it might be better to adopt the recently coined term 'biocatalyst' which aptly describes all of these situations.

It is beyond the scope of a text such as this to describe all the varied uses for which enzymes are at present employed. For a detailed review of the area the reader should refer to Godfrey and Reichert (1983). Three major steps can be taken between the formulation of a demand for a particular enzyme and its final production: source selection, extraction, and purification. Each of these will be dealt with as a separate topic but it must be remembered that in practice they are three steps of an integrated process. It will become apparent that the manner in which these three steps are tackled will depend upon a number of factors. For example, the source of the enzyme may affect both extraction and purification procedures; the final application of the enzyme will determine the degree of purification required; the scale of the operation may dictate which techniques can, in practice, be used.

### 13.2 Selection of sources of enzymes

Over 2000 enzymes have been isolated and characterized, the majority of them potentially available from any biological organism. The problem for the enzyme technologist lies in discerning from which source to isolate the desired enzyme. The solution is to a great part dictated by the characteristics of the desired enzyme, and of course the cost of the total process of isolation.

### 13.2.1 *Specificity*

If the desired enzyme is to be used in a process where a high degree of specificity is required, this may immediately limit the choice of the source. For example the first step in cheese-making, partial proteolysis of the milk casein to produce cheese curd, has traditionally used the enzyme rennin isolated from the stomach of suckling calves, clearly a limited and potentially expensive source. However, the high specificity of calf rennins leads to a very unique form of casein cleavage which is difficult to mimic with proteases from other sources. Some bacterial rennins have been isolated and used in cheese-making, but have not been universally accepted because they can, by producing subtly different proteolysis products from casein, produce off flavours in the final cheese. Conversely alkaline proteases used in detergent washing powders have low specificity of action, and substitution of one protease by another can be effective. In fact in this latter case, where the effect desired of the protease is to hydrolyze denatured insoluble proteins into small soluble peptides, mixtures of different proteases, each hydrolysing the protein into different peptides, can act synergistically. Thus far less total enzyme need be used when two proteases are added together, than when either one is used alone, in order to obtain the desired solubilization of the denatured protein stain in a given time.

### 13.2.2 *pH*

Consideration must be given to the pH at which the enzyme is to be used. Not only is the effect of pH on an enzyme's operational activity important but also the effect on the enzyme's stability. For example the animal protease trypsin is most stable at pH 3 but active only between pH 6 to pH 8; xylose isomerase (otherwise known as glucose isomerase) from *Bacillus coagulans* has an optimum range of activity of pH 5.5–7.0 but is operationally stable from pH 4.0 to 8.5. If the enzyme is to be of much use it must preferably be both stable and active at the necessary pH of the envisaged process. In addition the pH requirements of the upstream and downstream processes must be considered. For example, if the enzyme is active and stable at pH 7.0, and the upstream and downstream processes must be carried out at pH 10.5, then before the substrate is added to the enzyme considerable quantities of acid must be added and, after the reaction is complete, alkali added to return the pH to 10.5. The net product is a great deal of salt and additional expense. Far better to have an enzyme which works at pH 10.5. It was precisely this consideration that lead to the search for what is now the largest bulk enzyme, the alkaline proteases. Proteases have been in use as washing aids for over 50 years. Originally the enzymes were neutral proteases (that is, those stable and active at pH 7.0) which were incompatible with the typical pH of a washing powder detergent solution (approximately 10.5). Thus, such enzyme preparations had to be used in the form of a prewash soak. How much more convenient to have an enzyme that can be thrown in with the powder.

### 13.2.3 *Temperature*

That which is true of pH also applies to temperature. Contrary to the assertions of many a text book of biochemistry enzyme 'optimum temperature' is a myth. Figure 13.1 shows that the typical curve plotted for the response of enzyme's activity to temperature is composed of two antagonistic trends. Increased temperatures lead inexorably to an increase in reaction rate which is opposed by a denaturing effect of

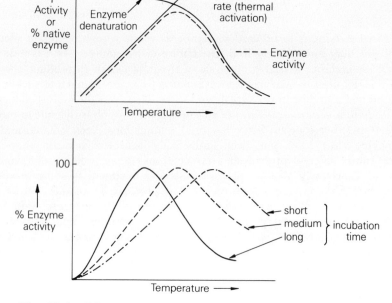

**Fig. 13.1**   Effect of temperature on enzyme-catalysed reactions

the increased temperature on the enzyme. However, while the increased reaction rate is an effectively instantaneous event, protein denaturation by heat takes a considerable time to occur. Thus at any given temperature the actual unit activity (in quantity of product produced per minute) will decrease as the total incubation time is increased. When comparing the ability of enzymes to act at a certain temperature care must be taken to ensure that the incubation time is the same for each enzyme or the result will be meaningless. In general, higher temperatures are desirable as processes will be less susceptible to microbial contamination and reaction rates will approximately double for every 10 °C rise in temperature thus speeding up the process. Thermal stability in the target enzyme may also be a useful attribute during the production of the enzyme itself as heat may be used to destroy contaminant enzyme activity.

Although the totally thermostable enzyme may be the ideal goal in the majority of cases, in some cases (for example, those using α-amylases) the enzyme may be required to be inactivated by heat at the end of the process.

### 13.2.4   *Activation/inhibition*
Similar enzymes isolated from different sources may require different activators or show different responses to a given inhibitor. For example bacterial β-galactosidases usually require cobalt as a cofactor; thus their use in removal of lactose from milk would entail the removal of the cobalt from the milk at the end of the process. Fortunately fungal β-galactosidases do not require cobalt. Removal of an activator from an enzyme may also alter its pH or thermal stability (see Section 14.5.4). Clearly the process for which the enzyme in question is destined, in particular the necessary required composition of the substrate or product may have a major influence on the choice of enzyme and its source.

13.2.5   *Analysis*
Little need be said except that appropriate analytical methods must be available to detect and quantify the enzyme and its products.

13.2.6   *Availability*
The availability of the enzyme and hence its source is of paramount importance. This is discussed below in further detail in Section 13.4. Not only must the enzyme be available but it and its source must be acceptable. This particularly applies when the enzyme is to be used in food processing or is to come into human contact (for example, washing powders). It is obviously more desirable to isolate a food enzyme from a non-pathogenic strain of organism!

13.2.7   *Cost*
The overriding consideration in industrial processing is that of cost. It is of little use to have a theoretically better enzyme if its cost is prohibitive. In this context cost relates both to the cost per unit of conversion and cost as a percentage of the total process cost. Such considerations are however less important in the field of therapeutic usage of enzymes or in areas of work such as genetic manipulation, where high purity and specificity are paramount.

13.2.8   *Stability*
Specific considerations of stability have been mentioned under the previous headings of pH and temperature above and will again be dealt with in the section on immobilized enzymes (see Sections 14.2, 14.5). However, a few pertinent observations can be made at this stage. The first is that the actual concentration of the enzyme may itself affect the stability of the preparation. For reasons which are not clearly understood, but may be due to protein aggregation, many enzymes are stabilized as their concentration is increased. Second, many enzymes may be stabilized by their substrates and/or products, for example amylases tend to be more stable in the presence of starch. Where an enzyme is stabilized by one of its products, inactivation, by, for example heat, at the end of the reaction may prove troublesome. Third, certain metal ions may stabilize enzymes. For example calcium ions at a concentration as low as 4 p.p.m will stabilize $\alpha$-amylase from *Bacillus licheniformis* to the extent that it retains 100% of its activity after 6 h incubation at pH 7.0 and 70 °C while in the absence of calcium it is completely denatured in 4 h under the same conditions. Curiously $\alpha$-amylase from *Bacillus amyloliquifaciens* is not stabilized by calcium. Fourth, enzymes may be stabilized by reducing the water concentration in the reaction mixture. For example, the liquefaction of starch can be successfully carried out using thermostable $\alpha$-amylase with a starch mixture at 80% dry solid at 110 °C! Some care must however be exercised as too low a water concentration may reduce the reaction rate and may lead to an altered product. $\beta$-galactosidase normally produces glucose and galactose from the hydrolysis of lactose in whey, but using concentrated whey the same enzyme will produce some glucose and galactose but also a mixture of trisaccharides (see also Section 15.3).

## 13.3   Sources of enzymes

Once we have defined the characteristics required of our chosen enzyme we may have already limited the potential source of the enzyme to one organism. However, it is more likely that we must still consider the relative merits of different sources of the same enzyme. In essence do we use a plant, animal or microbial source. Before we compare their relative virtues it is instructive to consider the sources of the most commonly utilized enzymes. The only major (in terms of volume of production) animal enzymes in use at present are trypsin, a variety of lipases and rennets; the traditional use of the enzyme content of dog faeces to remove hair from hides has lapsed in recent times. Plant enzymes are more prevalent and include the proteases papain, bromelain and ficin, cereal amylases, soyabean lipoxygenase and some specialized enzymes from citrus fruits. These enzymes are mostly used in the food processing industries. Microbial enzymes account for the rest (and by far the largest proportion by volume) but interestingly, despite the multitude of micro-organisms available fewer than 25 species (eight bacteria, four yeasts and 11 fungi) are actually used to produce virtually all of the microbial enzymes.The potential obviously exists to produce animal or plant enzymes in microbes by genetic transplantation but quite how such enzymes will be classified is open to question (would rennin produced from calf genes expressed in *E. coli* make vegetarian cheese?).

What is clear is that the source of an enzyme and its physical and chemical condition affects the efficiency of extraction, purification process, stability and ultimately cost of the enzyme. Up to the early 1970s it was considered that plant and animal material were the best sources of enzymes. Among the various reasons for such a view was the fact that the bulk of enzyme production was aimed at the food processing industry and enzymes isolated from plant or animal sources were deemed not to be tainted with problems of toxicity or contamination. After all if you can eat the beast you can surely eat its enzymes. Coupled to this, the relative ease of disruption of plant and especially animal tissue made enzyme extraction much simpler and less costly than from micro-organisms. However as the demand for enzymes grew, animal sources of enzymes became limited and expensive while plant sources were at the mercy of the weather and international politics, neither being particularly stable factors. This, together with the refinements occurring in fermentation technology, led to the rise in the fortunes of microbially produced proteins. It is perhaps significant that whereas in the last 5 years animal and plant-derived enzymes have risen considerably in cost in real terms, enzymes from microbial sources have actually decreased in cost in real terms.

## 13.4   Advantages of microbial enzymes

The obvious question to ask then is what has contributed to this change in relative costs; why are microbial enzymes commercially better? Microbial enzymes have two major types of advantage, technical and economic.

### 13.4.1   *Economic*
The first major economic advantage of microbial enzymes is that of bulk production. The sheer quantity of product that can be produced in a short time in a small area vastly exceeds that of animal or plant enzymes. For example a 1000 $dm^3$ fermenter of *B. subtilis* can produce up to 20 kg of enzyme in 12 hours. The average quantity of rennet extracted from one calf stomach is 10 g and it takes several months to produce one calf. Thus one gains from the economics of scale. The second advantage is that of ease of extraction. A large proportion of industrial enzymes, for example, most hydrolases, from microbial sources are produced extracellularly into the growth medium. Thus

there are no troublesome extraction problems. Even in the case of intracellular microbial enzymes extraction usually involves fewer steps than for plant and animal enzymes. For example the first steps in the production of enzymes from plant or animal enzymes involves the harvesting of the organism and its transportation to the processing plant. Micro-organisms are usually produced on the spot (with the notable exception of spent brewer's yeast) thus not only saving on the cost of harvesting and transport but also allowing the integration of the processes of production and extraction of the enzyme. Equally important is the consideration that the animal or plant enzyme, unlike that of the micro-organism, is usually specifically located in a particular tissue or organ and that this portion of the organism must first be separated from the rest and the remains disposed of. Two further economic advantages are shown by micro-organisms, the predictability of enzyme yield and the absence of seasonal variation. Plant and animal enzyme sources are subject to wide variations of yield and may be available only at certain times of the year. This latter problem therefore requires either the facility for high volume storage of the plant or animal source or else leaving extraction equipment idle for long periods of time, both factors contributing significantly to increase in cost.

### 13.4.2 *Technical*

There are four facets whereby microbes are more advantageous enzyme producers than plants or animals. The first is the enormous variation of biochemical pathways (and hence enzymes) that exist not only across the world of micro-organisms, but also within individual species. Thus one micro-organism is theoretically capable of producing a wide variety of different enzymes. The second advantage of micro-organisms is the range of environments in which they grow. This means particularly that enzymes stable under extreme environmental conditions are more likely to be found in micro-organisms which are adapted to growth in extreme environments. The most obvious example of this is the thermophilic organisms that often yield enzymes which are themselves stable at high temperatures. True thermophiles are considered to be those capable of growth at 65 °C; some organisms have been reported growing at much higher temperatures than this, for example *Methanothermus forvidus* at up to 97 °C and isolates from super-heated deep sea waters at 250 atmos. and 330 °C (see Baross and Demming, 1983). That is not to say that thermostable enzymes are not found in mesophilic organisms, indeed the protease thermolysin which retains 86% of its activity after 30 h at 70 °C is isolated from *Bacillus thermoproteolyticus* which itself is cultivated at an optimum temperature of 54 °C and cannot therefore be considered a thermophile. In structural terms the differences between an enzyme extracted from a thermophile and a mesophile are usually very small. However, these small differences endow the thermophilic enzyme with considerable advantages apart from the obvious one of eventual use at higher temperatures. For example, enzyme yield from thermophiles is often higher than from mesophiles because the enzyme is more resistant to conditions of extraction and purification, in particular the use of detergents or organic solvents. Equally, shelf life of the enzyme is usually better. Table 13.2 illustrates the difference in stability between glycerokinases from different sources.

The third advantage is the genetic flexibility of micro-organisms which results in the relative ease with which micro-organisms may be manipulated (see Sections 9.2, 9.3 and 11.2) to increase the yield of enzyme. The traditional techniques for yield improvement, those of strain or mutant selection, induction, derepression, or alteration of the growth medium are still preferred to gene transfer. This is not to imply that gene transfer does not have a great potential, rather that random mutation or

**Table 13.2**   Glycerokinase stability

| Temperature °C | Bacillus stearothermophillus | E. coli | Candida mycoderma |
|---|---|---|---|
| | | $t_\frac{1}{2}$ min | |
| 60 | 310 | 4.5 | 0.7 |
| 70 | 3 | 0 | 0 |
| 80 | 0.5 | 0 | 0 |
| 20 | no loss 10 days | 8.6 days | 4.1 days |

performance alteration of existing genes is a low cost technology, a rapid, proven method which does not depend on a (probably limited) knowledge of the genetics or biochemistry of a donor organism. The final advantage is the short generation times of micro-organisms. The average bacteria reproduces itself in a matter of minutes, whereas plants may take weeks and animals months to double their size. Even unicellular plants such as algae grown in suspension, have doubling times measured in hours. Thus starting with minimal quantities micro-organisms can produce in days what a plant may take months to yield.

Obviously future advances in plant and animal cell culture may reduce the significance of some of these advantages but even then where the desired enzyme is found only in a higher plant or animal, transfer of the appropriate genes to a rapid growing micro-organism may well be more cost-effective. However, while such generalizations and predictions may be valid where a single enzyme is desired, if the requirement is for multienzyme systems of plant or animal origin these considerations may change.

## 13.5   The problem of scale

Before we discuss the individual techniques of enzyme extraction and purification that are available to the enzyme technologist, some general consideration must be given to the problems of using such techniques in large scale operations. As a general truism it would be fair to say that what works best on a small scale will not work economically on a large scale. What follows is a discussion of the general problems of scale-up; specific problems will be considered with the discussion of individual techniques. The overriding considerations are those of yield and process time. Optimization of a process basically involves obtaining the highest yield for the shortest process time and in this optimizing procedure the following factors must be considered.

The larger the scale the longer it takes. The process of transferring a volume of liquid from one vessel to another is fast and simple if one has only 100 cm$^3$ to deal with; a quick tip and it is done. A volume of 1000 dm$^3$ requires plumbing, pumps and predictably takes a lot longer. Similarly heating a liquid from say 20 to 60 °C can be achieved rapidly for a small volume but may take some minutes for a large volume. Quite apart from the fact that the resulting increase in process time reduces the quantity of enzyme produced per unit time, such extended process times may reduce the yield because the enzyme will be held in adverse conditions (for example, of pH, presence of proteases, heat) for longer periods of time.

The physical process of handling large volumes is also a problem. The larger the process equipment the more complex it is liable to be, for example the need for pumping or cooling facilities generally will require more manpower to operate it. Thus increasing the process scale by a given factor leads to a disproportionate increase in cost which it is difficult to predict.

Temperature control is much more problematic on a large scale. Due consideration must be given to heating and cooling cycles when designing a process, and sufficient thought must be given to the requirement not only for heating but also refrigeration. Cooling is also a problem where large-scale cell disruption or centrifugation is undertaken. A bucket of ice around a $200 \, dm^3$ vessel is not really a feasible proposition. Not only, then, does the heating and cooling equipment add to the process cost but so does the power required to run them. Little wonder that using (warm) cooling water to heat greenhouses or fish tanks is a frequently employed method of offsetting costs. There is, however, a further problem relating to temperature control. On a small-scale process it is quite possible and usually desirable to control the temperature to within 1 °C. On a large scale the problem of inefficient liquid mixing and heat transference means that temperature variations of 10–15 °C are common. The danger is, of course, that this may lead to enzyme inactivation and hence a reduction in yield.

One of the major problems of scaling-up a process is that of equipment design where essentially two difficulties are encountered. The first is that no system behaves in an ideal and therefore predictable fashion. However, whereas on a small scale non-ideal behaviour may be insignificant, as scale is increased so the system deviates more from ideal behaviour. Thus, prediction of theoretical yields for a large-scale process from consideration of the small scale are usually overestimates, by an amount that in itself is unpredictable. The second problem is that processes that can be operated on a small scale often cannot be made to work on a large scale for simple engineering reasons. Centrifugation is a good example; ultracentrifuges capable of generating $150\,000 \, g$ are fairly commonplace objects in biochemistry laboratories. However, trying to build a large scale (that is $100 \, dm^3$ capacity) centrifuge capable of such forces by extension of the design of the laboratory centrifuge, would be akin to building Concorde by studying a Cessna. Similarly, chromatographic procedures must often be modified. The typical laboratory chromatography column is long and thin because that shape gives maximum product resolution. However, the compressible nature of the gels commonly used as chromatography media presents a mechanical problem if columns of more than a few metres in length are to be employed, because the pressure drop across the column packs the gel particles unevenly causing channelling or even blockage of the flow. There are two possible solutions to this problem: either design incompressible chromatography media, or redesign the column to make it short and fat. The latest generation of chromatography media to appear are significantly less compressible than their antecedents. However, despite the loss of resolution short, fat columns have an intrinsic economic advantage in that, for the same volume, less material and therefore less cost is required for their construction and the lower pressure drop across the column facilitates the use of lower power, less costly pumps. Another general design problem that must be taken into account is the desirability of continuous processes on a large scale. Thus batch centrifuges may be economically less viable on a large scale than continuous centrifuges, despite the latter's inherently lower gravitational force.

The next complication that must be considered is compatibility of equipment within the processing plant. On a small scale it makes practically no difference to economic considerations if equipment is dedicated to just one process. On a large scale, however, equipment must, at best, be multifunctional, capable of servicing more than one process. It is simply uneconomic to allow a large piece of equipment, which may represent a major capital investment, to lie idle for long periods of time. The different demands of different processes may therefore place constraints on the design of the equipment, resulting in a compromise non-optimal design for any of the processes. This may in turn reduce potential throughput or yield for some of the processes, but ultimately may make economic sense, because there is only the one capital cost to be offset against the final profits. Such considerations must also take into account the presence on-site of existing equipment and its possible suitability, as well as likely future use of any specially constructed equipment.

Finally, projected costs must be considered at the outset of the research and development programme. For example, the laboratory biochemist might develop a superbly elegant process based on affinity chromatography which, because of the cost of preparing affinity chromatography media would be hopelessly uneconomic on a large scale (p. 175). Even in the event that affinity chromatography must be used, there may well be economic sense in trading-off loss of resolution for cost of material, by using a cheaply available dye ligand (such as Procion dyes) in place of a more specific and costly 'biochemical' ligand. In addition, process cost is important not only in relation to product cost but also to product market volume. Thus, even though the target product may cost several thousand pounds per kg, a high technology high cost process will be a non-starter if the market volume is only a few hundred kg per annum. The total income from the product will be too small to repay the necessary investment.

## 13.6   Enzyme extraction

Once due consideration has been given to the selection of an appropriate source of enzyme the next stage is to extract it from the source. Numerous methods are available and choice of method will depend on the interaction of a number of factors, for example, nature of the source, scale of operation, stability of the enzyme, necessary purification. In this context large scale is commonly taken to mean starting with more than 1 kg of raw material although, quite obviously, for many of the enzymes produced in tonne quantities one is talking of many tonnes of raw material.

Enzyme extraction begins with an assessment of the actual cellular location of the enzyme: is it extracellular (for instance, many microbial-produced hydrolases) in which case no extraction is necessary, intracellular or even membrane bound?

When considering enzyme extraction procedures where some degree of cell disruption is required, there is great advantage to be gained by employing the least disruptive method consistent with the release of a major proportion of the required enzyme, thus keeping contamination of the extract to a minimum and simplifying subsequent purification. For example it may well be more economic to extract only 50% of the total available enzyme, with little contamination from other cell constituents in order to simplify and thus reduce the cost of purification. This point illustrates the need to consider the optimization of enzyme production as a whole, rather than optimizing each and every step individually.

Generally, soluble intracellular plant and animal enzymes present little extraction problem as the tissues are usually soft and may be easily mascerated by simple mincing or blending processes. Most micro-organisms are mechanically extremely robust and require more vigorous methods of disruption to release their soluble intracellular enzymes. Membrane-bound enzymes present a particular problem regardless of their source. The following is an account of a number of different methods of enzyme extraction and cell disruption, in practice mostly applied to micro-organisms, with an attempt to discuss their suitability for large-scale use.

### 13.6.1 *Abrasives*

It is perhaps no accident that one of the most commonly used methods of cell disruption is based upon the time-honoured technique of the pestle and mortar, the use of abrasives. On a small scale (1 dm$^3$) a laboratory blender supplemented with a handful of glass beads is effective at disrupting fairly tough microbial cells (such as the extraction of glucose dehydrogenase from *Bacillus subtilis*). On a larger scale various appliances operating on a continuous basis are available (for example, 'Dynomill'). Essentially they are all long cylinders, inlet and outlet at opposite ends with a longitudinal drive shaft to which are attached radial agitators (Fig. 13.2). Glass beads are placed in the container to provide the abrasive. Capacities commonly lie in the region of 0.6–15 dm$^3$. Many enzymes can be extracted from a variety of sources in such equipment, for example formate dehydrogenase from *Candida boidino* starting with more than 1 kg of cell paste. Rehacek and Schaefer (1977) have described a more efficient variant in which the agitators are placed alternately obliquely and radially along the drive shaft, and describe its use for disrupting yeast cells in a 20 dm$^3$ continuous-flow container. The system could handle 25–45 kg paste h$^{-1}$ as a 15% suspension at 150–250 dm$^3$ h$^{-1}$. They claim that their system could handle up to 340 kg h$^{-1}$ and had the advantage that it would disrupt streptococci, staphylococci, and micrococci, which are otherwise extremely resistant to disruption. Such equipment also has the advantage that it can be easily mounted in a safety cabinet, thus allowing its use with potentially pathogenic organisms, for example the extraction of restriction endonucleases from *Haemophilus spp.*

### 13.6.2 *Liquid shear*

The process of passing suspensions of cells at high pressure through a small orifice into a chamber at atmospheric pressure is known as liquid shear; the cells are literally blown apart by the sudden pressure drop. Such processes are of course amenable to continuous operation, and on a small scale pressures up to 200 MPa are used. Such sudden pressure changes, as anyone who owns a metal bicycle pump will know,

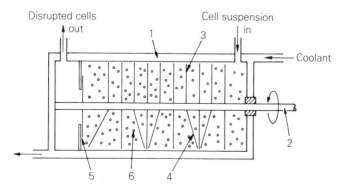

**Fig. 13.2** Principle of Dynomill type cell disruptor: (1) cooled drum; (2) rotating shaft (possibly cooled) to which discs are attached either (3) radially, or (4) alternately obliquely and radially; (5) baffle to prevent outflow of (6) glass ballotini

generate large quantities of heat; thus efficient cooling is required. Apparatus exists to handle up to $250 \, dm^3 \, h^{-1}$, however the larger the scale the lower the pressure drop and consequently the less disruption is achieved by a single pass. To overcome this reduction in disruption a recycling system may be used, but this of course offsets any advantages of the large scale and higher flow rate. This form of disruption is therefore one of those cases where the practical limitations of increased scale equipment may mean that large scale is not necessarily more productive. Liquid shear is, however a useful method for less robust organisms (for example, gram-negative bacteria) although the degree of disruption is often dependent upon the growth phase; stationary-phase cells proving more resistant than log phase cells. Robust gram-positive and fungal organisms may need to be continuously recycled in order to achieve maximal disruption.

### 13.6.3   *Osmotic shock*
The procedure of suspending cells in distilled water to release desired proteins is theoretically useful as the conditions involved are very mild. However, many organisms are quite resistant to osmotic shock and the process only finds real use with gram-negative organisms (such as *E. coli*) for the release of the enzymes, usually hydrolytic, of the periplasmic space. The virtue of this technique is that only these enzymes, which usually constitute less than 5% of the cell's protein, are released, thus simplifying purification. Some unusual enzymes may be released by this method, for example luciferase from *Photobacterium fischeri*. As a generally applicable method for large-scale operations osmotic shock is not ideal for three practical reasons, the large volumes of liquid involved ($400 \, dm^3$ for $10 \, kg$ cell paste), the large number of centrifugation steps involved and the necessity for low temperature.

### 13.6.4   *Alkali*
Perhaps the simplest method of cell lysis is treatment with alkali. Most cells will be disrupted at a pH between 11.5 and 12.5 and thus in theory alkalation of a cell suspension is a simple, cheap method easily applicable to a large scale. It does however suffer from one obvious drawback; the desired enzyme must be stable for 20–30 min at high pH. So far it has only been applied to the isolation of one enzyme on a large scale, asparaginase from *Erwinia chrysanthemi*.

### 13.6.5   *Detergents*
In theory again the use of detergents to disrupt cells is a widely applicable method. However, while most cells can be disrupted by detergents under certain conditions of pH, ionic strength and temperature, most enzymes will be denatured. In this respect ionic detergents are less preferable to non-ionic detergents as they tend to be more reactive, causing in particular lipoprotein dissociation. Quite apart from this problem the use of detergents complicates the process design and control, requiring accurate environmental control of the medium and ruling out certain useful purification methods, for example, salt precipitation. Despite this, detergents may prove useful in specific applications such as the extraction of membrane-bound enzyme, for example cholesterol oxidase from *Nocardia spp* is extracted using detergent, a process developed when other routine extraction methods failed. The process is described fully by Buckland *et al.* (1974) but essentially uses 0.5% Triton X-100 in $5 \, mmol \, dm^{-3}$ phosphate buffer at pH 7.5 to extract the enzyme, followed by batch adsorption and elution from DEAE Sephadex. This process had the added advantage that the protein

concentration in the supernatant was low (relatively little protein but all the required enzyme was extracted) and thus subsequent purification was simplified.

### 13.6.6   *Solid shear*
This method involves freezing a cell paste, usually to less than $-20\,°C$, and forcing it through a small hole at high pressure, and has traditionally been used as a batch, small-scale cell disruption process. Equipment for continuous processing of up to $10\,kg\,h^{-1}$ cell paste has been designed but is complicated and costly to operate and, with the exception of the isolation of heat sensitive enzymes, is unlikely to be widely applied.

### 13.6.7   *Lysozyme and EDTA*
The combined use of lysozyme (commercially prepared from hen egg white) to split bacterial cell walls and EDTA to lyse the cell membrane (by its chelation effect on calcium ions) is a delicate and specific method of cell disruption suitable for use on a small scale. Lysozyme alone is effective with gram-positive bacteria. However the high cost of lysozyme prohibits its use on a large scale.

### 13.6.8   *Organic solvents*
Although organic solvents are effective in disrupting a multitude of cell types and are a traditional method of cell disruption (for example, use of toluene), they are not used for large-scale enzyme extraction for a number of reasons: cost, toxicity, protein denaturation, flammability.

### 13.6.9   *Sonication*
Sonication is commonly used on a small scale to release intracellular enzymes. However, the high power requirement, the difficulty in transmitting the power to large volumes and the problem of heat dissipation make it impractical on a large scale.

## 13.7   Enzyme purification

### 13.7.1   *Introduction*
With one or two notable exceptions mentioned above, overcoming the problem of enzyme extraction generates the problem of enzyme purification. The process is usually divided into six stages: the removal of nucleic acids, the removal of cell debris, initial purification and concentration, final purification, concentration and packaging; but for the purposes of this discussion purification and concentration will be treated together as purification often leads to concentration.

However, before we begin our considerations, we should turn our attention once again to the ultimate criteria of large-scale enzyme production; is the process producing enzyme of the required degree of purity for the least possible cost? The obvious answer to this question might appear to be yes, so long as the integrated process has been fully optimized. However, one must bear in mind two factors which we have not yet discussed. First, most organisms are capable of yielding high quantities of more than one enzyme at a time; and second, there is usually a market for more than one enzyme. Thus it may be advantageous to consider the production and purification of more than one enzyme from one process. If this approach of multienzyme production is to be used two extra considerations become important. Compromises will inevitably have to be made to the optimal process for the production of any one enzyme, in order to allow the concomitant production of the second, third, fourth etc. enzymes. That is, the process as a whole must be optimized at the expense of optimum production of any one enzyme. Having two or more enzyme products from one process is clearly an economic advantage in that there is more product to pay for the cost of production, capital investment etc., always assuming that is, that the sum market value/volume of the different enzymes less their cost of production is greater than that for the single enzyme process.

An approach of this nature has been employed by Atkinson *et al.* (1979) to purify routinely four enzymes—superoxide dismutase, rhodanase, tyrosyl-tRNA synthetase and tryptophanyl-tRNA synthetase—from a single extraction of *Bacillus stearothermophilus*, a process which will also allow the recovery of 18 other enzymes (Fig. 13.3). In addition to the enzymes shown it was possible to adapt this process to produce the restriction endonuclease *Bst I*. A separate process to produce this enzyme would probably be totally uneconomic as it occurs at levels of only 0.004% of the soluble protein of the crude extract.

Finally, when undertaking enzyme purification, the general stratagem of the fewer the stages the less the loss, should not be far from the mind.

### 13.7.2 *Removal of nucleic acids*

Most enzyme extraction methods, with a few exceptions result in the more or less complete destruction of the cell. Thus not only is the desired enzyme released into the medium, but so too are the rest of the cell contents. A major component of most cells, and therefore in this context a major contaminant which must be removed, are the nucleic acids. Like most soluble polymers, nucleic acids add considerably to the viscosity of a solution which makes it more difficult to handle. It is for this reason that the first step of purification involves the removal of nucleic acids. Two approaches to the problem are available, precipitation and digestion. Precipitation of nucleic acids can be easily effected by the addition of high molecular weight polycations, for example protamine, streptomycin or polyethyleneimine. However, such polycations are relatively expensive, and are not usually used on economic grounds. Enzymic digestion with nucleases is, on the other hand, easy and cheap and for most applications is the method of choice. Clearly nuclease treatment does not actually remove the nucleic acids, it merely shortens them, but in so doing reduces the problematic viscosity of the mixture. The resulting oligonucleotides are removed in later stages of purification. The one area in which enzymic digestion is of little use is in the preparation of DNA ligases and exonucleases and endonucleases where high purity is required. Here nucleic acid precipitation must be employed, the extra cost adding to the premium price chargeable for such enzymes

### 13.7.3 *Removal of solids*

Once the nucleic acids have been precipitated or digested the insoluble debris of the cell must be removed. This debris is made up of precipitated nucleic acids, cell walls, large pieces of membrane, partially disrupted cells, etc. Care must obviously be exercised in not throwing the baby out with the bath water if the desired enzyme is membrane bound. Two approaches are possible, centrifugation or filtration, and the choice of method will depend in part on the desired scale of operation and in part on the nature of the solids.

On a moderate scale of up to 5 kg starting material it is often easiest to employ a large laboratory batch centrifuge. Centrifuges with capacities up to 6 dm$^3$ with forces up to 3000 g are commonly available. For larger scale separations (that is 5 kg starting

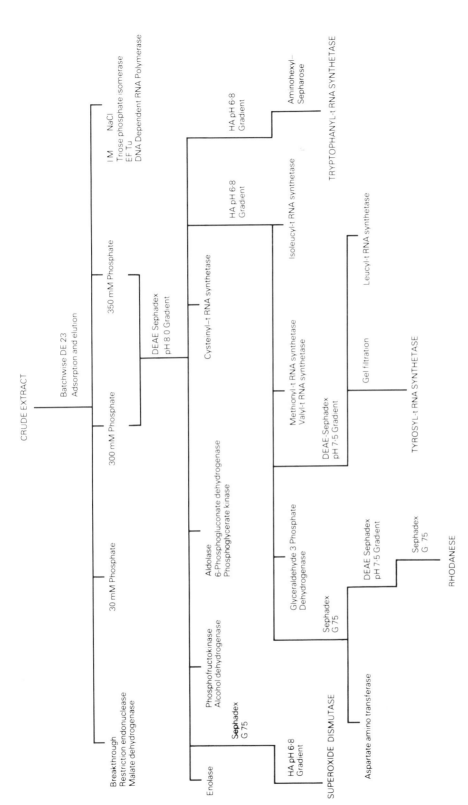

**Fig. 13.3** Flow diagram for a multienzyme preparation from *Bacillus stearothemophilus*. The four enzymes routinely obtained in homogeneous form are indicated in capitals and underlined. (Reproduced with permission from Bruton, 1983)

material) some form of continuous process will probably be desirable. Continuous centrifuges come in a wide variety of shapes and sizes, from disc type operating at 6000–8000 g retaining up to 60 kg of solid and handling several thousand $dm^3 h^{-1}$, to tubular types (up to 16 000 g), basket centrifuges (800 g) and even adapted domestic spin dryers! Some continuous centrifuges have the advantage of intermittent discharge of the accumulated solids which offers the operational advantage of not having to stop to empty the centrifuge. However, the discharge is usually a 50% slurry, thus some of the (enzyme) supernatant will be lost and economic sense must be utilized to ascertain whether intermittent discharge is cost effective (Fig. 13.4). While centrifugation is an excellent method for removing slimy or gelatinous solids (usually the case) which may be difficult to filter, it is not really useful with very large volumes of flocculate precipitates which may block centrifuge feed lines. In such circumstances filtration offers a better prospect.

### 13.7.4  *Purification and concentration*

The removal of nucleic acids and cell debris from the cell extract leaves a supernatant containing the desired enzyme. The next step in the process is the removal of unwanted contaminants, small inorganic and organic molecules, other proteins and most if not all of the water. A number of methods are available for the purification and concentration of the desired enzyme, some of which achieve both aims, some one or the other. The selection of the appropriate purification process is largely a mixture of empirism, experience and pragmatism and it is impossible to generalize. Table 13.3 summarizes a number of different enzyme extraction and purification procedures that have been used. It can be seen from examination of the data in this table that the total purification and concentration process is made up of a number of individual steps. First, an initial purification which may involve some concentration, followed by a further final purification usually involving chromatography, and finally concentration of the purified enzyme.

13.7.4.1  *Precipitation.* This is a time-honoured method of protein purification ideally suited to an initial purification of an enzyme extract. Precipitation may be of either a negative or positive type, that is either the impurities may be precipitated or the target enzyme. The latter is perhaps preferable because the precipitated enzyme may then be redissolved in a minimal volume thus achieving concentration of the enzyme. A number of different precipitants may be used. Adjusting the pH of the extract from 7.0 down to 5.0 or 4.0 will often remove soluble cell wall components and many undesired proteins; it has been used in the initial purification of nucleoside-5'-phosphotransferase from carrot by Marutzky et al. (1974). Obviously this method is only of use where the desired enzyme is stable and soluble at low values of pH. Perhaps the most widely used precipitant is ammonium sulphate, the use of which in fractional precipitation procedures is an example of a mixture of positive and negative precipitation—incremental addition of solid ammonium sulphate initially precipitating unwanted contaminant proteins then finally precipitating the desired enzyme. Quite considerable purification and concentration can be achieved by simple and low cost routines, which perhaps explains the ubiquitous nature of this technique.

In addition, high ammonium sulphate concentrations can quite considerably increase the stability of some enzymes.

On a very large scale however ammonium sulphate presents handling and disposal problems as at high concentration it will corrode concrete and even the high quality stainless steel used in the construction of large-scale facilities. Sodium sulphate is not

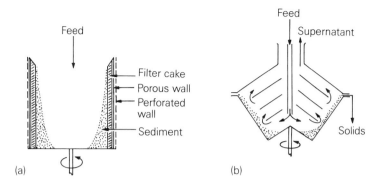

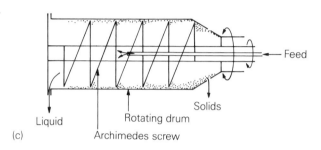

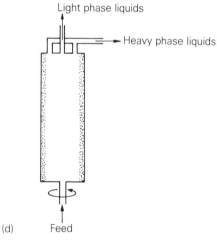

**Fig. 13.4** Types of continuous centrifuge; (a) basket; (b) disc bowl; (c) solid-bowl scroll; (d) 'Sharples Super'

corrosive, but its lower solubility necessitates the use of temperatures above 35 °C to achieve the same degree of fractional precipitation.

Organic solvents, while effective precipitants suitable for laboratory work, are rarely used on a large scale because of their flammability. Various polymers may be used as precipitants. Polyethylene glycol (PEG), which is both non-toxic and non-flammable, is used in serum fractionation; its use is restricted, however, because

## Table 13.3   Methods of enzyme purification

| Enzyme | Source organism & purification method | Reference |
|---|---|---|
| L-Asparaginase | *Citrobacter* spp. (55 kg paste) | Bascomb *et al.* (1975) |
| | Manton Gaulin homogenizer | |
| | MnCl$_2$ precipitation of nucleic acid | |
| | Acetone fractionation | |
| | Ammonium sulphate fractionation | |
| | 8 dm$^3$ DEAE-cellulose column with gradient elution | |
| | 2 dm$^3$ hydroxyapatite column with gradient elution | |
| | 3 × 1.4 dm$^3$ Sephadex G200 columns in series | |
| $\beta$-Galactosidase | *Escherichia coli* | Robinson *et al.* (1974) |
| | Manton Gaulin homogenizer | |
| | Ammonium sulphate fractionation | |
| | Affinity chromatography on $p$-aminophenyl-$\beta$-D-thiogalactoside-agarose column (1.8 dm$^3$), eluted with 0.05 M phosphate pH 7.0 and 0.1 M borate pH 10.0 | |
| Formaldehyde and formate dehydrogenases | *Candida boidini* (1 kg paste) | Schutte *et al.* (1976) |
| | Dynomill operated at 5 dm$^3$/h | |
| | Streptomycin sulphate to precipitate nucleic acid | |
| | 7.8 dm$^3$ DEAE-cellulose column with step elution | |
| | 1 dm$^3$ DEAE-cellulose column with gradient elution | |
| | 1 dm$^3$ hydroxyapatite column with gradient elution | |
| Pullulanase and 1,4-$\alpha$-glucan phosphorylase | *Klebsiella pneumoniae* (5 kg paste) | Hustedt *et al.* (1978) |
| | Extract cells with sodium cholate | |
| | Two liquid phase partition using 9% PEG and 2% dextran | |

| | *Upper phase* | *Lower phase* |
|---|---|---|
| | Remove PEG by diafiltration | Manton Gaulin homogeniser |
| | Precipitate PULLULANASE | Add PEG to 12% and repartition |
| | with cetyl-trimethyl ammon- | Add PEG to 19% and KCl to |
| | ium bromide | 1.22 M to upper phase |
| | | Take lower phase |
| | | Remove PEG by diafitration |
| | | Batch adsorption and elution with |
| | | CM-Sephadex of 1,4-$\alpha$-GLUCAN |
| | | PHOSPHORYLASE |

| Enzyme | Source organism & purification method | Reference |
|---|---|---|
| Renin | Pig kidney (100 kg) | Hue *et al.* (1976) |
| | Mince tissue | |
| | Ammonium sulphate fractionation | |
| | Affinity chromatography on 0.5 dm$^3$ column of pepstatin-Sepharose | |
| | Gel chromatography on Ultrogel AcA 44 | |

of its high viscosity in solution. Polyacrylic acid, also non-toxic, has been used in the partial purification of amyloglucosidase from *Aspergillus niger*. One general problem of precipitation methods is that they are batch processes which are difficult to integrate into continuous processes.

13.7.4.2 *Adsorption.* Adsorption onto insoluble materials, in batch processes, has been much used as an early stage in the purification and concentration of proteins. Three main types of material can be used: resins, substituted dextrans or agarose, and substituted celluloses. The adsorption principle used is usually that of ion-exchange, but with some materials, for example cellite or hydroxyapatite, adsorption is non-specific. Ion exchange resins such as sulphonated polystyrenes are, in practice, of little use for the adsorption of proteins, because their high degree of substitution can lead to denaturation. The exception to this is their use with proteins with unusually high isoelectric points such as cytochrome *c*. Substituted celluloses, agaroses or dextrans, functioning as ion-exchange materials are much used, although on a large scale they tend to be costly. Materials such as cellite and hydroxyapatite, although they are relatively non-specific in their interaction with proteins, are useful because of their low cost. Hydroxyapatite in particular is useful, as a differential elution of the many proteins bound can be achieved with increasing concentrations of phosphate. In addition, chloride is not a counter-ion and thus adsorption and elution can be carried out in up to 1.0 M chloride. Affinity type materials are rarely used at this stage because of cost considerations. For all of the above materials processing of the crude extract is usually performed by batch adsorption and elution rather than as column chromatography. There are various reasons for this which are concerned with the large volumes of low clarity extracts to be processed. First there is the problem of the compressibility of the adsorption material. Most ion-exchange, or for that matter gel filtration, materials based on celluloses or agarose tend to be rather squashy gel materials. Consequently, when placed in a column, if too high a pressure is applied to the buffer flowing through the column the gel material will be compressed to the point where channelling of the flow will occur, or worse still the column will become blocked. The only effective way of reducing the pressure is to reduce the flow rate, construct very short, fat columns, or run a number of short columns in series. This latter point is discussed below. Short fat columns tend not to have the high resolution characteristics of long thin columns. The maximum flow rate achievable on laboratory style chromatography equipment is about $30 \, cm^3 \, h^{-1}$, which for processing several hundred $dm^3$ is not really fast enough! The low degree of extract clarification also complicates the use of column chromatography, because the top layer of material in a column acts as a very efficient and easily blocked filter! Consequently, where large volumes of murky material are to be handled, batch processing is the realistic answer despite the relatively low degree of resolution. Typically adsorbant will be stirred into the extract for a short while, and spun off in a basket centrifuge (or even the spin dryer). The adsorbant would then be resuspended in a medium of higher ionic strength, altered pH, or both and then spun down once again. Using a centrifuge with a capacity of up to $40 \, dm^3$ it is quite feasible to process up to $2000 \, dm^3 \, day^{-1}$. During the elution process it is more efficient to divide the eluent up into two or three lots rather than to use it all at once. As a compromise batch adsorption with column elution may be employed. In summary the net gain by using such processes is time saved, the net loss is resolution.

13.7.4.3 *Two liquid phase separation.* This is a relatively recent introduction into the protein chemist's armoury. It has two great potential advantages, easy separation of proteins from bacterial cell walls and semisoluble debris, and its conversion to a continuous process. The principle of the technique lies in the observation that when different proteins are placed in a solution of two immiscible polymers, the two polymers will separate into distinct phases and the proteins will be selectively adsorbed into one or other phase. The reason for this partitioning of polymers into phases and selective adsorption of proteins lies in the physical nature of the molecules involved; identical or similar molecules have a greater tendency to aggregate with each other rather than with non-identical or dissimilar molecules. Typically, the two-phase system might use a mixture of polyethylene glycol and dextran or perhaps polyethylene glycol and ammonium sulphate. In which phase the protein ends up depends upon the molecular weight of the protein, concentrations and molecular weights of the polymers, temperature, pH and ionic strength. Selection of the appropriate conditions is entirely empirical. Once the ingredients have been mixed, separation into the phases takes place by settling or centrifugation (possibly continuous). Various examples of this technique have been quoted; one such is the separation and purification of pullulanase (pullulan-6-glucan hydrolase) and 1,4-α-glucan phosphorylase from *Klebsiella pneumoniae* (Hustedt *et al.*, 1978, see also Kula, 1979). Cell debris was separated from the pullulanase in a polyethylene glycol–dextran mixture, the polymers removed by combined dialysis and filtration and the enzyme precipitated with N-cetyl N,N,N-trimethyl ammonium bromide to give 70% yield at 80% purity. The cell debris and phosphorylase in the initial dextran phase were then separated in a different polyethylene glycol dextran system.

13.7.4.4 *Column chromatography*. This is usually reserved for the later stages of processing once a degree of purification and concentration (hence reduction in volume) has been achieved. All types of chromatography normally used in the laboratory can be employed although gel filtration, ion exchange and affinity chromatography are the major methods. The older type of chromatographic materials (e.g. Sephadex, DEAE Sephadex, etc.) are gradually being superseded by later generations based either on crosslinked celluloses and agaroses or polyacrylamide/agarose copolymers, or on composite inorganic/cellulose materials. All of these materials are designed to give more rigid, smaller particles and can thus be used at high flow rates and yet give greater resolution. This greater inherent resolution may be employed to economic benefit by shortening the columns, thus reducing both the cost of the column and its contents. While a shorter column will not actually speed up output rate (obviously controlled solely by the flow rate) it will have two further practical advantages; it will reduce both the pressure drop across the column, and the unit operational time of the process; that is, because of the smaller total volume of the system it will take less time to pass a given volume liquid through the column. Clearly there has to be a disadvantage to these newer chromatography techniques, and not surprisingly it is found in their cost.

The problem of large-scale chromatography does not end, however, with the introduction of materials with better hydrodynamic properties. For example, the difficulty of packing a $10 \, dm^3$ column is more than that of packing a $500 \, cm^3$ column, fraction collection at several $dm^3$ a time requires special apparatus. Although columns are available with capacities up to $100 \, dm^3$, achieving even and predictable flow through such a column is difficult. It is preferable therefore to use ten $10 \, dm^3$ columns connected in series, whence the whole system will have flow characteristics of a single $10 \, dm^3$ column.

Gel filtration, particularly when using one of the newer media, can be a very efficient method of initially separating the desired enzyme from low molecular mass molecules. Flow rates of the order of $2.0 \, dm^3 \, h^{-1}$ through a $2.5 \times 80 \, cm$ column are possible (compared to $30 \, cm^3 \, h^{-1}$ for 'standard' material).

Ion exchange chromatography is often used in conjunction with or in place of gel filtration, and generally has a higher resolution than gel filtration. Ion exchange also allows for the useful possibility of batch adsorption followed by column elution, which, when employed with the newer generation matrices with their high flow rates, can considerably shorten process time. However, due to their cost these newer matrices are not greatly used on a large scale, despite their advantages (Table 13.4).

**Table 13.4** Summary of the advantages of crosslinked substituted agarose

---

1. Incompressibility = high flow rates
2. High capacity even at high ionic strength
3. Do not shrink/swell on changing pH/ionic strength
4. Regenerate in column
5. Autoclavable
6. Short cycle times

---

Affinity chromatography, at least in the classical sense, does not find much use in large-scale operations. Despite the advantages of high resolution and one or two step purification processes, the expense, low capacity and instability of the affinity matrix usually rule out its use except for the removal of contaminating enzymes or where the final enzyme product cost is unimportant (for example, some diagnostic enzymes).

Affinity chromatography can be divided into two types—specific and general—according to the nature of the affinity ligand. Specific ligands are usually substrates, their analogues or inhibitors of the enzyme and as such usually demonstrate high specificity and binding energy. General ligands include molecules such as AMP, hydrocarbons (absorbing enzymes by quasi specific hydrophobic bonding) or dyes. There is extensive literature available in this area and for details of some applications the reader is referred to the works of Robinson *et al.* (1974), Hue *et al.* (1976) and Furbish (1977). In general, it will suffice to record for the purposes of this discussion that affinity techniques often shown an increase of yield (over conventional alternatives) from 5% to 50–90% and typically may reduce the complete purification process to three or four steps.

It is, however, worth elaborating a little on the growing use of the triazine dyes (such as Procion blue) as affinity ligands. This group of compounds have a number of advantages over traditional affinity ligands: low cost, ease of immobilization to the matrix, and stability of the matrix–ligand band. The triazine dyes have a lower degree of specificity than traditional ligands, but by judicious use of conditions, dye type or eluting molecule, sufficient specificity of binding can be achieved for all practical purposes. It is not known precisely how these dyes bind to their target enzymes, but presumably it must be a result of similarity of shape/size to the enzymes, but presumably it must be a result of similarity of shape/size to the enzyme's substrate, coenzyme etc. This particular branch of affinity chromatography promises to have a great future in enzyme purification and has been reviewed by Lowe (1981).

13.7.4.5 *Continuous electrophoresis.* The practicality of using the electrophoretic properties of proteins to purify enzymes (or proteins) on a preparative scale was low until the invention of continuous electrophoresis. The principle involves the application of an electrical potential across a continuous stream of protein solution an entirely obvious idea. However, because of sideways mixing across the stream in practice the technique will not work, until laminar flow is obtained; that is some method of preventing the sideways mixing. Like so many useful discoveries the principle of the induction of laminar flow is extremely simple. The electrophoresis 'device' is composed of two electrodes: a cylindrical solid cathode, surrounding which is a hollow cylindrical anode that is made to rotate slowly. The solvent is passed between the two electrodes, and the rotation of the outer electrode swirls the solvent around the inner electrode stabilizing the flow and preventing mixing across the gap between the electrodes (Fig. 13.5). Fluid flow rates of up to 1 dm$^3$ h$^{-1}$ are possible with the solvent collected at 20 separate ports at the top of the device. This technique is still in its infancy (and the hardware still expensive), and at present is mostly used for the fractionation of blood proteins. However, its potential for the separation of isoenzymes and virtually pure mixtures of enzymes as well as particulate materials cannot be ignored.

13.7.4.6 *Ultrafiltration and dialysis*. The application of semipermeable membranes to large-scale enzyme purification is largely confined to latter stages of the process and for concentration and salt removal. In this context dialysis is impractical because of the large volumes of buffer/distilled water which would be required. Ultrafiltration on the other hand can be used very effectively to concentrate protein solutions. Two configurations of ultrafilter exist, the hollow fibre and flat bed filter. Hollow fibre ultrafilters are simple to apply and a moderate size unit can handle 50 dm$^3$ h$^{-1}$ with ease. However they tend to bind the filtered proteins if their concentration falls below about 1 mg cm$^{-3}$. Flat bed filters have less tendency to bind proteins, but generally (because of the lower available surface area achievable in practice) have lower flow rates (for example, 4 dm$^3$ h$^{-1}$). Both types of filter can be difficult to clean and some enzymes are inactivated by the shearing forces produced at the membrane surface. However, recent advances in ultrafilter design and manufacture are permitting ultrafiltration to become a key operation in downstream processing, both for protein concentration and size separation, being preferred to other techniques because of low cost and ease of maintaining sterility.

### 13.7.5  *Concentration and packaging*

The extent and nature of final concentration and packaging of an enzyme will depend both on what is practicable, the use to which the enzyme is to be put, and whether large quantities of the enzyme must be transported over long distances. The ultimate consideration is to perform the least possible number of processes in order to obtain the enzyme in an acceptable form. For enzymes for analytical or small-scale investigative manipulative work the enzyme is usually packaged either as a lyophilized

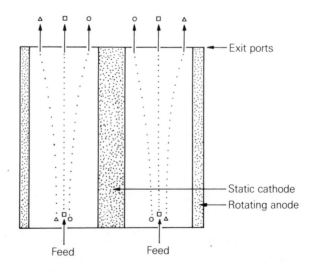

**Fig. 13.5**  Principle of continuous electrophoresis

powder (from freeze drying) or as a suspension in 3.4 M ammonium sulphate. If the purified enzyme is to be further immobilized on site it may be possible to leave it as a relatively pure but weak solution. For enzymes required in bulk, particularly proteases for detergents, some form of packaging in a dust-free form must be employed for reasons of both safe handling and ease of transport. In these cases the enzyme is mixed with various salts and waxes and formed into pellets, or else supplied as a concentrated solution. Where complete drying of an enzyme is required and freeze drying is not practicable because of the scale or instability of the enzyme, then evaporation can be employed. Heat denatures proteins, but only if it is applied for a long enough period (usually >0.5 s). A device is manufactured by Alfa-Laval, called the Centi-Therm, which effectively evaporates enzyme solutions to dryness using this principle of ultra-short heating time and evapourative cooling. The heater is a stack of hollow, conical, heated discs rotating on a common spindle. Enzyme solution is sprayed onto the underside of the discs, the spinning motion of which spreads the liquid to a layer 0.1 mm thick. The liquid is thus in contact with the heated discs for less than 1 s, boils more or less instantly and the dried enzyme is flung clear of the heater.

## 13.8   Summary

Enzymes are potentially of great use because of both their enormous catalytic activity and high degree of specificity. Enzymes may be isolated from almost any organism, but the selection of a source for enzyme production will depend upon the consideration of a number of factors—for example, specificity, pH dependence, heat stability, activation or inhibition, availability and cost. In practical terms, the use of micro-organisms to produce enzymes has a number of technical and economic advantages, and in recent years has become the predominant mode of enzyme production.

The large-scale extraction and purification of enzymes poses a number of problems not met at the laboratory scale, and thus not all potential methods are suitable. In particular the nature of the whole process must be known before attempts can be made to optimize any one step, as the choice of any one particular method for a given step in the process may influence the suitability of other methods to be used. The emergence in recent years of novel techniques of enzyme extraction and purification promises markedly to improve such processes.

# Chapter 14

## Enzyme applications

### 14.1 Introduction

#### 14.1.1 *Areas of use*

Now we have managed to isolate and purify the enzyme of our choice, we can move on to consider the practical applications of enzymes and the problems (and hopefully solutions) of these applications.

There is, as we shall see, an enormous range of applications of enzymes, all however unified by those uniquely enzymic characteristics of highly efficient catalysis under mild reaction conditions coupled to very precise specificity for both substrate and reaction catalysed. A cursory inspection of the storeroom of almost any industry or activity associated with organic chemical or biological concern will usually reveal the presence of enzymes. Thus 'enzymes' can be found in both likely (such as hospitals) and unlikely (such as the production of acrylamide and oil fields) places. For the sake of clarity we shall here divide our discussion into four categories of usage: developmental, therapeutic, analytical and industrial processes.

#### 14.1.2 *Usable catalysts*

However, before we consider each of these fields in detail we should elaborate a little on what makes an enzyme a usable catalyst.

14.1.2.1 *Catalytic efficiency.* The catalytic efficiency of enzymes is superbly demonstrated by the enzyme catalase. This enzyme catalyses the breakdown of hydrogen peroxide into oxygen and water. It does so by virtue of a single atom of iron located in the active site of every molecule. There is nothing particularly magic about this iron atom. In fact iron filings are capable of catalysing the decomposition of hydrogen peroxide. However, what might have appeared magic to our ancient alchemists is the observation that the rate of breakdown of hydrogen peroxide achieved by 1 mg of iron in catalase is equal to that achieved by several tonnes of iron fillings. This difference in catalytic efficiency can be attributed to the environment provided for the iron by the enzyme. Thus the turnover number of most enzymes (that is, the number of molecules of substrate which 1 molecule of enzyme can process per second) falls within the range $1-10^6$. Some of the most efficient enzymes, for example acetyl cholinesterase, with turnover numbers around $10^5 s^{-1}$ are in effect limited not so much by their catalytic ability but by the speed (frequency) with which enzyme and substrate molecules collide. As we shall see (Section 14.2) this can have consequences in the catalytic behaviour of an enzyme.

14.1.2.2 *Specificity*. Enzymes in general are highly specific in terms of both substrate utilized and reaction performed. Catalase will only break down hydrogen peroxide into water and oxygen; iron filings will, however, catalyse a large number of different chemical reactions. Such specificity is, of course, invaluable enabling enzymes to carry out very precise conversions, picking out of a mixture of chemicals just one molecular type—hence the value of enzymes in the analysis of complex mixtures of chemicals. However, this high degree of specificity does have some drawbacks in terms of their practical application. For example, the alkaline proteases, such as subtilisin or thermolysin, which are added to washing powders to help degrade proteinaceous stains, are superb at hydrolysing haemoglobin, but not nearly as effective with casein or albumin; if you must stain your clothes do it with blood not milk or eggs! Another curious example of this problem of 'specificity' can be found in the production of high fructose corn syrups (see Section 14.7). In this process glucose is isomerized to fructose by the enzyme xylose isomerase. Unfortunately, at least for the purveyors of cheese whey wastes, the xylose isomerase will not perform the same trick on galactose, nor has an enzyme yet been found which will isomerase galactose to glucose. Such a discovery would enable the conversion of an essentially waste product, whey, into a highly valuable sweetener, fructose syrup.

14.1.2.3 *Stability*. It is perhaps self-evident that to be of much practical use an enzyme must be stable, and must retain its catalytic ability for the time required to carry out the desired reaction. Thus the enzyme must be relatively resistant to any denaturing agents, chemical or physical, found in the reaction mixture. However enzyme stability is important for a number of other reasons. First, a stable enzyme is going to survive any adverse conditions encountered during isolation more easily, giving a higher yield and therefore cheaper product. Second, stability enhances shelf-life and ease of storage of a product. Third, the more stable an enzyme the greater the potential for reuse (see below). Fourth, in many instances it is advantageous to run enzyme reactions at elevated temperatures, for example to speed up the reaction and to avoid problems of microbial contamination. Aspects of stability of enzymes are discussed at greater length in Sections 14.2.3 and 14.5.

Finally, however, one should remember that there are occasions where highly stable enzymes are a disadvantage, notably in processes such as the saccharification of starch where a soluble enzyme must be removed from the product stream. The easiest way of achieving this is by heating the mixture.

14.1.2.4 *Reusability*. With a few notable exceptions enzymes are not cheap. In order to lessen the cost of an enzyme-based process it is of great advantage if the enzyme can be used more than once. The greatest problem with enzyme reuse lies in the recovery of the enzyme from the reaction mixture. Theoretically it is possible to extract the enzyme from the reaction mixture; perhaps the simplest process might be ultrafiltration, but the cost is usually prohibitive. For this reason much of enzyme technology is directed towards the problem of reuse of enzymes. The most successful approach has been that of enzyme immobilization (see Section 14.2). Reuse of an enzyme implies either that the enzyme must be retained in a continuous stream of substrate, or else it must be separated/removed from the final reaction mixture.

14.1.2.5 *Removal from reaction*. Removal of the enzyme from the product of the reaction is often the first step in the purification of the product. If the enzyme is cheap enough and reuse is not required the simplest method is to heat the mixture and filter off the denatured enzyme. Sometimes the product cannot be heated by the required amount without producing deleterious effects (this applies particularly in food processing), and so some other method of protein removal must be used, for example, ultrafiltration, precipitation by change in pH or ionic strength.

### 14.1.3   *Catalyst enhancement*

Clearly no enzyme can be an ideal catalyst largely because the factors which contribute to the usability of the enzyme will differ in importance depending on the particular use envisaged. Thus stability may be the prime consideration if an enzyme is to be used as an industrial catalyst, whereas specificity may be more important if that same enzyme is required as an analytical tool. However, there are a number of ways and means by which an enzyme's useful characteristics may be enhanced and it is the purpose of this section briefly to introduce those methods. Further discussion of these methods occurs in other sections of this book.

14.1.3.1   *Selection.* In theory at least, the simplest way of improving an enzyme's characteristics is to select the appropriate source material. Thus if thermostability is the overriding consideration it makes sense to search for the enzyme among thermophilic or caldoactive organisms. In addition, isoenzymes from one source organism often show quite distinct characteristics. Where specificity is important again appropriate selection of source organism may prove useful. For example, calf rennins for cheese-making are highly specific, and any attempt to replace them with rennins from microbial sources must take this specificity into account.

14.1.3.2   *Reaction environment.* Frequently it is possible to manipulate the reaction environment to enhance an enzyme's useful characteristics. Thus amylases (and many other enzymes) are more stable in high substrate concentrations.

Most enzyme reactions are, in theory at least, reversible. For example, the hydrolysis of peptides by proteases is a reversible reaction whose equilibrium under normal dilute aqueous conditions lies far to the right (that is in favour of hydrolysis). This is largely a result of the high water concentration. Consequently, it is in practice possible to reverse this reaction by lowering the water content of the reaction mixture by the addition of an organic solvent or, alternatively, by continually removing the peptide product. Such approaches are used in the synthesis of the dipeptide sweetener aspartame (see Section 15.3, and Carrea, 1984).

14.1.3.3   *Chemical modification.* Chemical modification of an enyzme can produce an amazing array of changes in the enzyme's characteristics, from alterations in pH optima and $K_m$ to enhancement of stability (see Section 14.5). The approach tends to be very empirical, and all too often attempts at chemical modification result in enzyme inactivation. As a technique for 'mass production' of enzymes however it has distinct limitations because of the potentially high costs.

14.1.3.4   *Enzyme engineering.* The term enzyme engineering has recently been coined to apply to the modification of enzyme structure by alteration of the enzyme's gene. Most of the attention in this field is at present devoted to the use of site-directed mutagenesis to produce single amino acid substitutions in the enzyme's primary structure. Potentially, at least, it should be possible to alter any characteristic of the enzyme in this way. However, to be successful, it is necessary to have a complete knowledge of the structural basis of the particular characteristic to be changed and what other characteristics might also be altered in the process. Thus it would be pointless changing the structure of the enzyme in order to decrease the $K_m$ if in so doing the enzyme's stability is markedly reduced. At this time such structural information is not readily available for most enzymes. Indeed little is known of the general features which lead to enzyme stability (see Section 11.4 and Winter and Fersht, 1984).

Quite apart from making minor modifications to an enzyme's structure it may also prove possible to design whole new protein molecules by joining together the genes for distinct enzyme or protein molecules, to create a new hybrid enzyme. These new multigenes could then be expressed to give a multienzyme molecule. It is now clear that this happens widely *in vivo*; fatty acid synthetase is such a molecule, one peptide chain containing three globular domains each with its own different enzyme activity. The potential therefore exists for building multienzymes for the catalysis of whole metabolic sequences with all the operational advantages of a multienzyme complex (see Section 10.1.3). One exciting prospect utilizing this technique is the attempt to fuse together an enzyme such as peroxidase and protein A (from *Staph. aureus*) for use as an immunoglobulin-binding enzyme in enzyme linked immunosorbent assays (ELISA; see Section 14.7.2).

14.1.3.5 *Immobilization.* If every there was a goose with a golden egg in enzyme technology then enzyme immobilization must surely be the prime candidate. However, opinion seems divided as to the precise date of ovulation. That immobilization has been hailed as the 'cure-all' for the problems of enzyme technology would be to understate the case. Certainly the immobilization of an enzyme makes recovery and reuse of the enzyme easy, may improve the enzyme's stability and all too often alters kinetic characteristics such as pH optimum, $K_m$, sensitivity to inhibition, although not necessarily in the desired fashion. However, despite all this promise, immobilized enzymes are found in relatively few applications: approximately seven major industrial processes, a few suggestions for therapeutic use and a handful of routine analytical applications. Why should this be? The answer lies in two words: cost and effectiveness. Although immobilized enzymes are easier to handle than their soluble counterparts, they are often not such effective catalysts in that they introduce into industrial processes almost as many problems as they solve. Equally the cost of immobilization may itself render their use uneconomic. These points will be elaborated on in the discussion of enzyme applications and immobilization which follow in the rest of this chapter. Is there then a future for immobilized enzymes? The answer is undoubtedly yes, but depends above all else on unpredictable economic factors. However, the trends are perhaps emerging. Greater use is beginning to be made of immobilized cells as biocatalysts and it may be this field in which future promise lies (see Sections 14.4 and 14.7). Perhaps in the immediate future the greatest potential of immobilized biocatalysts lies in the development of biosensors, the immobilization of enzyme or cell around a sensor device capable of responding to the catalyst's action. The sensor might be in the form of a potentiometric electrode (such as a pH electrode) or even a silicon chip. This latter concept of a stable bioelectronic sensor, if it eventually proves feasible, would have exciting applications in a range of fields from medicine to industrial processing, as it could be used not just for monitoring but could be directly incorporated into control systems.

## 14.2 Immobilization

### 14.2.1 *What is immobilization?*

Immobilization of an enzyme has been defined as 'the imprisonment of an enzyme in a distinct phase that allows exchange with, but is separated from, the bulk phase in which substrate, effector or inhibitor molecules are dispersed and monitored' (Trevan, 1980; Figure 14.1). This definition may be extended to cover cells and other biocatalysts. The enzyme phase is usually a water-insoluble high molecular weight polymer (such as cellulose), and 'imprisonment' of the enzyme may be achieved by a variety of means. The potential usefulness of such a technique has been outlined above (Section 14.1.3) and will be discussed further (Section 14.5, 14.7). First, however, we must consider briefly the means by which biocatalysts may be immobilized and the effect such immobilization has upon the biocatalyst's behaviour.

### 14.2.2   *Methods of immobilization*

Immobilization methods may be subdivided into various groups, depending upon the physical relationship of the catalyst to the polymer matrix. Thus the catalyst may be covalently bonded to the polymer, physically adsorbed onto the polymer, crosslinked with itself (and possibly another inert protein), entrapped inside a polymer matrix or encapsulated in a 'polymer bag'. Any combination of these methods may also be used. Figure 14.2 illustrates these five major categories. Several hundred methods of immobilization have been described, some more generally applicable than others, and many more are doubtless yet to be discovered. However, despite this diversity of method there are relatively few methods in general use, and it is on these we shall concentrate. In all cases the term polymer or polymer matrix is used loosely to imply the material which effectively immobilizes the enzyme; almost any material imaginable may be employed from cellulose hydrogels to materials such as nylon, glass or even iron filings. The reader is referred to Woodward (1985), for fuller accounts of such methods.

14.2.2.1   *Adsorption*. Adsorption of an enzyme or cell to a polymer material probably represents the earliest method of immobilization. A range of non-specific or specific bonding forces may be employed, for example electrostatic, hydrophobic interactions or affinity bonding to specific ligands attached to the polymer. Tate and Lyle instituted an immobilized invertase process in the 1940s based on a packed bed of invertase adsorbed onto activated charcoal (which also served to remove contaminating coloration from the syrup). Perhaps one of the most widely used immobilization processes is that of electrostatic bonding of the catalyst to an ion exchange cellulose (such as DEAE cellulose). This type of method has a number of potential advantages: ease and mild conditions of preparation of the immobilized material (simply stir polymer and enzyme together); availability of ready prepared polymer material suitable for use in column reactors; potential for regeneration of the immobilized catalyst; often little limitation of access of substrate to the bound enzyme; application

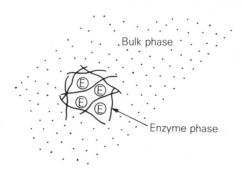

**Fig. 14.1**   Schematic definition of enzyme immobilization

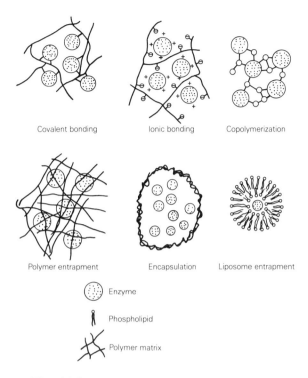

Covalent bonding    Ionic bonding    Copolymerization

Polymer entrapment    Encapsulation    Liposome entrapment

Enzyme

Phospholipid

Polymer matrix

**Fig. 14.2**  Modes of enzyme immobilization

to whole cells or organelles. However, there are two serious drawbacks: leaching of enzyme or cells occurs relatively easily with changes in pH or raised ionic strength, sometimes simply by an exposure of the enzyme to its substrate; substrates carrying the same charge as the polymer may be prevented, by virtue of ionic partitioning (see Section 14.2.4), from gaining access to the enzymes except at high concentration levels which may in turn cause loss of enzyme from the polymer matrix.

14.2.2.2  *Covalent bonding.* Many varied forms of covalently bonded immobilized enzyme preparations have been reported. Of this multitude a handful of methods are in routine use. In principle, there are two ways of covalently bonding an enzyme to a polymer, the first by activating the polymer with a reactive group, the second by the use of a bifunctional reagent to bridge between enzyme and polymer. The principal groups on the enzyme through which it is coupled are hydroxyl and amino groups and, to a lesser extent, sulphydryl groups. This raises one of the major problems with this technique, that the enzyme may often be inactivated by virtue of conformational change consequent upon reaction, or reaction with a group at the enzyme's active site. This latter problem can often be overcome, however, by performing the immobilization in the presence of the enzyme's substrate or a competitive inhibitor, or, with proteases, using the zymogen form (see also Section 14.5). Among the commonest forms of activated polymer are hydrogels, for example celluloses or polyacrylamides, onto which have been incorporated diazo, carbodiimide or azide groups. Alternatively such hydrogels may be directly activated with, for example, cyanogen bromide. This last technique in particular has found wide application. The use of bifunctional reagents is an alternative strategy for covalent enzyme immobilization. Here two approaches are possible. The simplest, using for example glutaraldehyde, involves mixing together polymer, enzyme and reagent simultaneously, a simple although not very controlled process. The second approach is to react the bifunctional group with the polymer to the desired degree of substitution and then add the enzyme. The advantage of this approach is that the bifunctional reagent is effectively prevented from penetrating the enzyme's active site. One multifunctional reagent used in this way is cyanuric chloride (trichlorotriazine), which is particularly interesting because it is a trifunctional reagent, the reactivity of its groups becoming lower as it becomes more substituted. Thus pure trichlorotriazine will react with cellulose in a matter of seconds to produce cellulose-dichlorotriazine, which can be reacted with a hydroxy-substituted compound, for example, ethanolamine, in a matter of minutes to give cellulose-monochlorotriazinylethanolamine, which in its turn will take several hours to react completely with an added enzyme. By using a compound of the type X—OH to perform the second substitution, positively, negatively or uncharged groups may be introduced into the polymer to alter its ionic or hydrophilic/hydrophobic properties, either to enhance attraction of the enzyme to be bound, or to alter the properties of the final immobilized enzyme preparation.

Included in this category of immobilization method is the use of metal ion complexes to link catalyst and polymer. For example, titanium (IV) ions form extremely strong (although strictly non-covalent) multiple interactions, and a number of examples exist of the use of titanium (IV)-coated glass used as an activated polymer matrix for binding enzymes and cells. It is worthy of note that the use of metal ion complexes is one of the few 'covalent' bonding methods usable with cell preparations.

The general advantages of covalent bonding in the preparation of immobilized catalysts are similar to those described above for adsorption, with the notable variations that leaching of the catalyst is generally not a problem; the polymer can be constructed to carry any sign or degree of charged groups; almost any polymer material may be used; the variety of chemistry available means almost any enzyme is potentially immobilizable; the ability to fix enzymes to soluble polymers. Among the disadvantages of this method must be included: the frequent inactivation of the enzyme; the use of toxic reagents; often complicated preparative routines; the general inability to yield immobilized cell (as opposed to unpurified enzyme) preparations.

14.2.2.3 *Crosslinking.* Crosslinking of the enzyme to itself by reaction with a bifunctional reagent, with the possible inclusion of an inert protein, is essentially an extension of covalent bonding techniques with all the potential advantages and problems that this brings. The bifunctional reagent most often used is glutaraldehyde. The technique is cheap and simple but is not often used with pure proteins because it yields very little bulk of immobilized enzyme with a very high intrinsic activity. It is, however, widely used in commercial preparations (such as glucose isomerase; see Section 14.7.4) of immobilized enzymes derived from non-viable cells or crude cell extracts, largely because of its low cost.

14.2.2.4 *Entrapment.* Entrapment of biocatalysts within a polymer matrix is in principle easy to perform. The biocatalyst is dissolved in a solution of the polymer's precursors and polymerization initiated. Two types of polymer have found widest application, polyacrylamide type gels and naturally derived gel materials such as cellulose triacetate, agar, gelatin, carrageenan or alginate. The major advantages of this type of method are the simplicity of, and mild conditions used in, the preparation of the immobilized biocatalyst and the usefulness of this method for immobilizing cell preparations. However, because of the need to keep the average pore size of the gel as high as possible to prevent excessive diffusion limitation (see Section 14.2.4) and the variability of pore size in such gels, there may be great leakage of the biocatalyst from the gel, particularly with low molecular weight enzymes. Non-viable cells are usually retained well, but viable, dividing cells may burst from the gel material (Fig. 14.3). None of the entrapment methods commonly used is perfect. Thus, polyacrylamide gels suffer the drawback that the monomers from which they are formed and the free radicals generated during polymerization are toxic, therefore excluding this method when using viable cells or sensitive enzymes. Agar and carrageenan have large pore sizes allowing quite sizeable ($<10\,\mu$) cells to escape easily; in addition, they are depolymerized by mild heat. Calcium alginate gels, to date the method of choice for viable cell immobilization, are disrupted by calcium chelating agents such as citrate or phosphate which may often be necessarily included in the reaction medium.

One novel variation of this technique involves the preparation of fibres of immobilized enzyme. Pastore and Morisi (1976) spun fibres of $\beta$-galactosidase entrapped in cellulose triacetate for use in the hydrolysis of lactose in milk, and more recently a technique of spinning fibres of calcium aginate entrapped cells has been patented. The particular advantage of these fibre preparations is their relatively high surface area to volume ratio and thus reduced diffusion limitation (see Section 14.2.4).

14.2.2.5 *Encapsulation*. Encapsulation of biocatalysts by enclosing a droplet of solution of the biocatalyst in a semipermeable membrane capsule has to date largely been restricted to medical applications. The technique is simple and cheap, but to be effective the biocatalyst must be stable in solution. Relatively little diffusion limitation is encountered and capsules of defined size can be easily formed by alteration of the preparation conditions. Materials used to form the capsule may be 'permanent' (for example, nylon, or biodegradable for example, polylactic acid or phospholipid

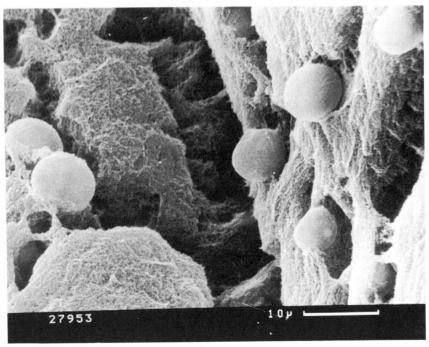

**Fig. 14.3** Scanning electron micrograph (SEM) of algal cells immobilized within a calcium alginate gel matrix (photo by C. W. Lilley and M. J. Taylor)

liposomes). Although the catalyst (be it cell or enzyme) is very effectively retained within the capsule, such preparations are mechanically unstable. Another variant of this form of immobilization is the entrapment of a solution of enzyme in the lumen of hollow, semipermeable fibres of the type used in membrane filtration cartridges. The fibres are bathed in a stream of substrate which permeates into the lumen, reaction occurs and product diffuses back out of the lumen into the substrate stream. The advantages of this technique are that it gives an immobilized enzyme preparation with a very high surface area to volume ratio, is easy to perform, allows easy replacement of the enzyme, and is usable with high viscosity or particle-containing substrate streams which might block reactors based on other forms of immobilization. Its major disadvantages is that the enzyme is in no way protected in the sense that there is no potential for stabilization of the enzyme, as can occur with other methods. Indeed, adsorption of the enzyme onto the walls of the fibre may be accompanied by denaturation. The application of this technique is also limited by the molecular weight of the substrate.

14.2.2.6   *Choice.* Choice of immobilization method is discussed below (see Section 14.6.2) and it will suffice to record at this point that it is based largely upon empirical factors. Few general rules exist, although some more obvious considerations (such as substrate molecular weight, size and inherent stability of enzyme molecule, requirement for a specific physical form of the immobilized enzyme) may rule out certain methods. There is usually, however, little alternative to the try-it-and-see principle. The choice of method for the immobilization of amino acylase (see Section 14.7.4) was a result of a 'screening' of over 40 different methods.

### 14.2.3   *Effect on stability*

The stability, and methods for enhancing it, of enzyme or cell preparations, a major concern of biocatalyst technology, is dealt with in several places in this book (see Sections 14.1.2, 14.1.3, 14.5). A few words must be said here, also, on the effect of immobilization on enzyme stability.

While there are many real ways in which immobilization may stabilize an enzyme (see Section 14.5), many of the reports of stability induced by immobilization are probably the result of limitation of free access of the substrate to the enzyme. Thus although enhanced 'operational' stability may be observed, no intrinsic stabilization is achieved. Figure 14.4 is a schematic representation of the results often obtained from stability studies and is a consequence of what has been called the 'Zulu factor'. Zulu warriors were an effective fighting force not just because of their dedication and military skills but because of their sheer weight of numbers. When the first wave of warriors were killed there would be another to take its place, so to the attacked the ferocity of the battle would be maintained despite the high death rate of the Zulu force. So too with immobilized enzymes. If the immobilized enzyme particle has a high loading of enzyme, substrate molecules diffusing into the particle meet a very high local enzyme concentration and thus may penetrate only the outer 10% or so of the particle before being totally converted into product. With time the denaturation of a fraction of the enzyme occurs. However, there may still be more than sufficient enzyme present to allow the substrate to penetrate the particle to any extent, and thus the effective catalytic activity of the particle remains constant. This effective catalytic activity is a result of the limitation of free diffusion of the substrate and is equal to the diffusion rate, which may be considerably lower than the intrinsic enzyme activity of the particle, thus loss of some of the enzyme will have little effect upon the reaction rate. At high substrate concentrations the rate of diffusion of substrate is increased and

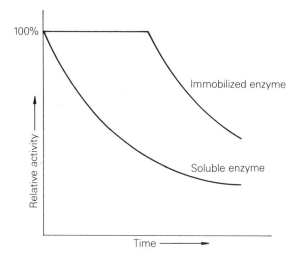

**Fig. 14.4** Schematic illustration of apparent stabilization of an enzyme by immobilization

may exceed the intrinsic activity of the enzyme with the result that this observed form of stabilization is lost. Alternatively, with time, sufficient of the enzyme may become denatured so that the point is reached where enzyme activity, rather than substrate diffusion, is rate-limiting.

Whether this effect is real or illusory it can be put to good use, as it gives a preparation with operational stability, that is the ability to maintain a constant level of activity over a significant period of time during actual use. This results in easier process control and in the linearity of response of many enzyme-based biosensors (see Section 14.7.2).

The same principle can apply to both enzyme and cell preparations, but with the latter other factors may be involved. Certainly many reports exist of the stabilization of cells by immobilization. This may be due to the provision of a more suitable environment for the cell by the polymer matrix (for example, local concentration of endogenous hormones), or may be the result of real stabilization of the cells' membrane. For example, many claims have been made for the stabilization of plant cells by immobilization, cell life extended by a factor of 10 or more. Here, however, the stability may be a direct result of the experimental comparison used. The longevity of immobilized plant cells has usually been compared to that of the same plant cell grown in suspension culture. However, the immobilized cells have the physical characteristics of callus tissue which is eminently more stable than free cell suspensions.

Perhaps the most intriguing report of activity stabilization in cells is that by Takata *et al.* (1982), who demonstrated that different methods of immobilization had varying effect on the stability of fumarase activity in *Brevibacterium flavum* cells. It should be noted that no indication was given of the precise metabolic state of the *B. flavum* cells and there is no clear way of telling whether they were viable cells, whole non-viable cells or a lysed cell/crude enzyme preparation; the use of $1 \, \text{mol dm}^{-3}$ fumarate and bile salts as a pretreatment of the cells make the latter most likely. In essence their studies showed that the stability of κ-carrageenan entrapped cells was some five times greater than that of polyacrylamide entrapped cells and ten times greater when the polycation, polyethyleneimine was included in the carrageenan gel. Polyethyleneimine was shown to have some stabilizing effect upon the fumarase activity of the free cells, and also to enhance the stability of polyacrylamide entrapped cells when included in their preparation. Takata *et al.* (1982) concluded that this stabilization (to treatment by heat, urea, high pH and ethanol) was a result of a three-way interaction between the cell membrane, the carrageenan and the polyethyleneimine. However, it must be added in caution that the immobilized preparations had relatively high loading of cells (16% by weight) which might suggest that some of the 'stabilization' observed was due to diffusion limitation. In addition, the data presented in this and previous reports showed (a) some of the hallmarks of diffusion limitation (for example, broadening of pH profiles, less than expected rise in activity with increased temperature etc.); (b) changes in physical characteristics of the carrageenan gel with inclusion of polyethyleneimine. It seems possible, therefore, that the differential stabilization observed could be due to interaction of various factors; (a) some genuine stabilization of activity of the cells *per se* by polyethyleneimine (polyamines are used as cryoprotectants); (b) variations in gel structure resulting in gels of differing diffusional resistance (the inclusion of starch in carrageenan gel entrapped cells lowered their stability); (c) different immobilization methods causing different levels of enzyme inactivation during preparation, leading to different initial loadings of activity (acrylamide monomers and the free radicals produced during polymerization are known to be toxic).

The reason we have dwelt a little on this tale is that it neatly illustrates the difficulty in ascertaining precisely the causes of apparent biocatalyst stabilization. It must be clearly pointed out, however, that in practice when a biocatalyst is to be put to real use, the cause of superior stability is of secondary importance to the fact that stability has been enhanced. He cares not of the means who by the end profits.

In conclusion it should be clear that great care must be exercised in claiming superior intrinsic stability for a biocatalyst once immobilized (even when stabilization may be apparent) as real stabilization is probably the exception rather than the rule.

### 14.2.4    *Perturbation of kinetic characteristics*
Much has been written on the effect of immobilization on the kinetic characteristics of biocatalysts, most of it relating to enzymes rather than cells. We will thus content ourselves with a brief review of the more salient features which may cause the kinetic characteristics to be perturbed.

14.2.4.1    *Characteristics considered.* In the following discussion we shall confine ourselves to a consideration of the following, $K_m$, $V_{max}$, the effect of pH, the effect of inhibitors. In this discussion some terminology new to the reader may be encountered particularly in relation to the terms $K_m$ and $V_{max}$. Enzymes very often demonstrate changes in $K_m$ and $V_{max}$ as a consequence of immobilization and a number of notations have been used to distinguish between the $K_m$ and $V_{max}$ of the free enzyme and those measured for the immobilized enzyme. Here, when considering the immobilized enzyme, we shall use the terms $V_s$ and $K_v$ for the maximum attainable velocity under saturating substrate conditions and the substrate concentration giving a velocity of $\frac{1}{2}V_s$, respectively.

14.2.4.2    *Causes of perturbations.* There are two reasons why immobilization may result in an (apparent) change in an enzyme's characteristics. Immobilization may directly affect the enzyme molecule's structure and hence function, or it may provide a microenvironment around the enzyme different from that of the bulk phase. The microenvironment provided by the polymer matrix may affect the enzyme in one or both of two ways, by partitioning molecules between the bulk phase and the enzyme's environment, by restricting the free diffusion of molecules into or within the polymer matrix. Any or all of these effects may be present and may be summarized thus:

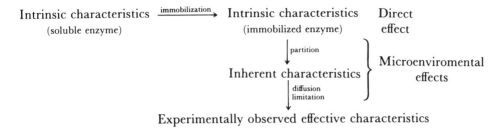

Direct effects of immobilization on an enzyme's intrinsic characteristics could include total or partial inactivation caused by gross conformational change or reaction of some essential group at the enzyme's active site; more subtle induced conformational change causing destablization, alteration of allosteric effects or kinetic characteristics; stabilization (see Section 14.2.3 and 14.5).

The microenvironment surrounding an immobilized enzyme, provided by the polymer matrix, may affect the enzyme's apparent behaviour by causing heterogeneous distribution of solutes between the enzyme and the bulk phase in which the immobilized enzyme is dispersed. These microenvironmental effects may be subdivided into two categories, partitioning and diffusion limitation (Fig. 14.5). Partitioning results from hydrophobic or electrostatic interactions between the polymer matrix and the solutes present, and affects reacting and non-reacting solutes alike causing their concentration or depletion in the vicinity of the enzyme.

Limitation of the free diffusion of solute molecules by the physical presence of the polymer matrix will cause the depletion of substrate molecules and the concentration of product molecules around the enzyme. Diffusion limitation may be of two types. The barrier to diffusion may be the unstirred layer (Nernst layer) of solvent surrounding the immobilized enzyme particle, so-called external diffusion limitation, and in this case reaction occurs subsequent to diffusion. Alternatively (or as well)

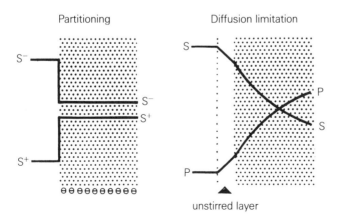

**Fig. 14.5**  The effect of partitioning and diffusion limitation on solute concentration profiles in immobilized enzyme systems

diffusion limitation may occur within the polymer matrix, internal diffusion, and reaction and diffusion are occurring concurrently. Where diffusion limitation occurs up to or in a reacting immobilized enzyme particle, a steady state is eventually reached where the rate of diffusion of a solute into or from a given volume element of the particle will equal its rate of removal or production by the reaction. To those with a mathematical mind the use of terms such as 'volume element' might imply the necessity for the use of calculus in the quantitative modelling of immobilized enzymes and, for the unmathematically minded, this is unfortunately the case. However, we will not here attempt to treat such effects quantitatively, but instead to derive a qualitative, intuitive understanding.

14.2.4.3  *pH effects*. Extensive study of the literature on the subject of the effect of pH on the activity of an immobilized enzyme, compared to the effect on the same enzyme in its soluble native form, will reveal all manner of differences. The pH 'profile' of the enzyme (that is, a graphical plot of rate versus pH) may be displaced, distorted, broadened or narrowed as a consequence of immobilization. How are we to discern the cause? Briefly there are four potential causes which may exist individually or in combination; partitioning of hydrogen ions; electrostatic charge on the polymer; limitation of diffusion of hydrogen ions produced or consumed by the reaction; substrate diffusion limitation.

*Partitioning* of hydrogen ions towards or away from polyionic matrix supports will result in concentration or depletion of hydrogen ions in the vicinity of the enzyme. Thus the pH around the enzyme may be lower or higher than that measured in the bulk phase and this will lead to an apparent shift of the pH profile to higher or lower pH values (Fig. 14.6). The extent of this wholesale shift of the pH profile will depend upon the partition coefficient for hydrogen ions of the polymer matrix ($\Delta H^+$) which will be $>1$ where the polymer is an anion and $<1$ for a cation. $\Delta H^+$ is defined as the ratio of the hydrogen ion concentration at the polymer surface ($H_i^+$) to the concentration in the bulk phase ($H_e^+$) (that is, $H_i^+/H_e^+$). For any absolute value of the ($H_i^+$) this ratio will remain constant and thus so will the pH difference between the polymer surface and the bulk phase ($pH_e - pH_i$). Like all ionic partition coefficients, $\Delta H^+$ is dependent upon ionic strength and will be reduced to near unity by high buffer or salt concentrations. For example, it has been observed that an enzyme immobilized onto a polyanionic matrix (negatively charged) may show an apparently raised pH optimum (a pH profile displaced to higher values). This may be explained by considering that, in this case, the polymer is concentrating hydrogen ions, thus lowering the pH, around the enzyme and therefore in order to obtain the pH value for optimum activity in the vicinity of the enzyme, the external bulk phase value of pH must be higher than the enzyme's intrinsic pH optimum. The same argument will apply for any given point on the enzyme's pH profile, and because for a given value of $\Delta H^+$ the pH difference ($pH_e - pH_i$) remains constant regardless of the absolute values of $pH_e$ or $pH_i$, the profile will appear to be shifted without distortion towards more alkaline values of pH.

*Electrostatic charge on the polymer* may have a direct effect on the stability of ionizing groups at the enzyme's active site thus raising or lowering their pKa value. For example, trypsin is dependent for its activity on the presence of an ionized serine hydroxyl group, which has a much lower pKa value when serine is present in trypsin than when it is present in dilute solution. This lowering of the pKa value, in effect stabilization of the $-O^-$ ion, has been postulated to be due in part at least to the net positive charge of the surface of the enzyme molecule (6+ per molecule). Modification of this surface charge to a net 5− raises the pH optimum of the enzyme by 1 pH unit,

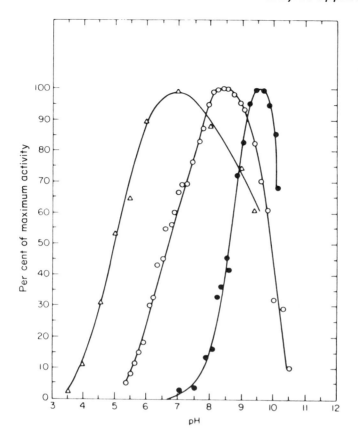

**Fig. 14.6** Effect of the polymer matrix on the pH profile of chymotrypsin, in dilute solution (○), immobilized to polyanionic ethylene–maleic anhydride copolymer (●), and immobilized to polycationic polyornithine (△). Substrate acetyl-L-tyrosine ethyl ester, ionic strength 0.01. (Reprinted with permission from Goldstein, 1972)

while modification to a net 12+ lowers the pH optimum by 1 pH unit. Both of these effects are a result in the change in stability (and hence pKa) of the serine $- 0^-$ ion. Obviously the same sort of explanation could be invoked to explain the effect of polyionic matrices on the pH profiles immobilized enzymes. In practice it is impossible to distinguish between this effect and the partitioning of hydrogen ions described above.

*Diffusion limitation* of hydrogen ions, despite their high diffusional mobility, by a polymer matrix may well occur when substrate or product diffusion limitation is not seen, because of the relatively low concentration of hydrogen ions encountered in most systems. Here, as in other cases, diffusion limitation will only be apparent if hydrogen ions are consumed or produced by the reaction. It so happens that this is the rule of enzyme reactions rather than the exception.

Let us take the example of a reaction producing hydrogen ions, where the enzyme shows a typical bell-shaped pH profile and is immobilized at an impervious surface. This latter assumption simplifies matters slightly, because we will only be concerned with external diffusion limitation and we can effectively think of the system as two homogenous compartments, the bulk phase and the enzyme's microenvironment. As with partitioning effects any difference between the value of $pH_i$ and $pH_e$ will give rise to a displacement of the pH profile of the immobilized enzyme, because the enzyme will be operating at a different pH to that measured. Regardless of pH, once the reaction is started, hydrogen ions will be produced in the microenvironment and, in order for them to diffuse into the bulk phase, they will accumulate until a sufficiently great concentration gradient is produced. Ultimately the rate of diffusion out of the microenvironment must equal the rate of production. Of course this is not the end of the matter, for hydrogen ions have a marked effect upon enzyme activity, and what happens next depends upon whether the initial pH is above or below the pH optimum. Above the pH optimum the accumulation of hydrogen ions will cause the microenvironmental pH to fall, which in turn raises the enzyme reaction rate. This results in an increased rate of production of hydrogen ions which causes greater accumulation, forcing the pH down further still, the rate increases more, more hydrogen ions accumulate, this autocatalytic cycle continuing until the further accumulation of hydrogen ions and drop in pH produces a negligible increase in reaction rate. Thus the microenvironmental pH is 'backed off' towards the pH optimum, the effect being observed as a broadening distortion of the alkaline limb of the pH profile (Fig. 14.7). If the initial pH is below the pH optimum, then the initial accumulation of hydrogen ions (and lowering of $pH_i$) will reduce the rate of reaction, lessening the need for accumulation, the effect becoming self-limiting, such that in this region the microenvironmental pH will not differ significantly from the bulk phase pH and the acid limbs

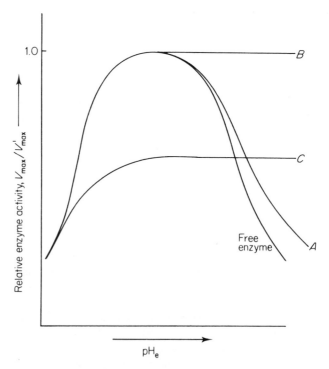

**Fig. 14.7**   Schematic representation of the effect of $H^+$ ion diffusion limitation on $H^+$ ion liberating enzyme immobilized onto ionically neutral polymer. $(A)$ mild, $(B)$ moderate, and $(C)$ severe diffusion limitation (Reproduced with permission from Trevan, 1980)

of the immobilized enzyme will appear little changed from that of the free, soluble enzyme. The extent of such distortions will depend on two factors, the severity of diffusion limitation and the intrinsic enzyme activity. The greater the enzyme activity the more apparent diffusion limitation will become. Where there is very severe diffusion limitation the microenvironmental pH may never exceed the pH optimum (curve B, Fig. 14.7), or indeed may not even reach it (curve C, Fig. 14.7). This last suggestion may at first seem illogical; however, it must be noted that the rate of diffusion of hydrogen ions from the microenvironment will be proportional to the concentration difference $H_i^+ - H_e^+$. In cases of severe diffusion limitation a large difference is required such that $H_i^+ - H_e^+$ approximates to $H_i^+$. This approximation is valid where the $H_e^+$ is less than 1% of $H_i^+$, that is, $pH_e - pH_i > 2$. Thus we may have the situation where, in order to provide a sufficient gradient to drive the efflux of hydrogen ions at the same rate as their production, a hydrogen ion concentration difference of $10^{-6}$ mol dm$^{-3}$ is required and thus the value of $pH_i$ cannot exceed 6 (as at this value $H_e^+$ must be effectively zero). Clearly if the intrinsic pH optimum of the enzyme is greater than 6, the immobilized enzyme will always operate at $pH_i$ values below the pH optimum.

It should be noted that this type of effect is most likely to be present in the absence of buffering solutes, whose presence may remove any potential pH gradient.

Finally, one must take account of the fact that the logarithmic nature of the pH scale means that the same difference $H_i^+ - H_e^+$ will be reflected as an increased difference $pH_e - pH_i$ as the absolute value of $H_i^+$ is raised. This in itself will lead to an increased distortion of the pH profile with increased pH, giving a broadening of the alkaline limb of the pH profile.

*Substrate diffusion limitation* may also affect the response of an enzyme to changes in pH. Typically the pH profile of the enzyme will be broadened, the enzyme showing a decreased sensitivity to pH changes. Fig. 14.8 is typical of this type of effect, which can be explained by considering that, when substrate diffusion limitation is severe it effectively controls the rate of reaction and is of course unaffected by changes in pH. At the extremes of the range of pH over which the enzyme is normally active, the intrinsic activity of the enzyme falls below the rate of substrate diffusion, thus becoming the controlling factor in the effective rate of reaction, which of course now responds to changes in pH.

*Combinations* of all these various effects can occur with the result that under a particular set of conditions the response of the immobilized enzyme to pH can be quite extraordinary. Figures 14.9 and 14.10 are two such examples.

14.2.4.4  *Michaelis constant* $(K_m)$. Changes in the $K_m$ observed for an immobilized enzyme are the rule rather than the exception. Leaving aside for the moment any direct effect of immobilization on the intrinsic $K_m$ of the enzyme, let us consider those effects which may be brought about by the microenvironment of the immobilized enzyme. Here again two factors may contribute to the microenvironment, partitioning and diffusion limitation.

*Partitioning* of the substrate towards or away from the polymer will obviously result in a substrate concentration surrounding the enzyme different to that measured in the bulk phase. Let us assume that the substrate partition coefficient

$$p = \frac{S_i}{S_e}$$

where $S_i$ is the substrate concentration at the polymer surface and $S_e$ that of the bulk

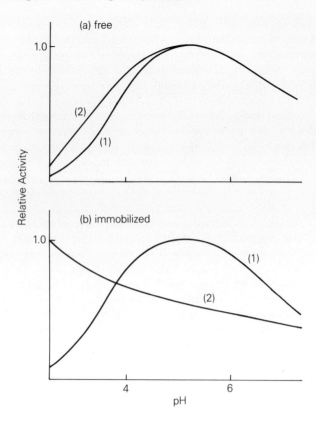

**Fig. 14.8** The effect of substrate diffusion limitation on the pH profile of polyacrylamide gel entrapped glucose oxidase. (a) Free enzyme in the presence of (1) 1.0 and (2) 0.1 mol dm$^{-3}$ glucose; (b) immobilized in the presence of (1) 1.0 and (2) 0.1 mol dm$^{-3}$ glucose. Note that the effect, like diffusion limitation, vanishes at high glucose concentrations. (From M. D. Trevan, unpublished data)

phase. For a simple system where the enzyme obeys Michaelis–Menten kinetics, the rate of reaction ($v$) will be related to the value of $S_e$ and $p$ in the following way

$$v = \frac{V_{max}S_i}{K_m + S_i} = \frac{V_{max}pS_e}{K_m + pS_e}$$

which, dividing by $p$, becomes

$$v = \frac{V_{max}S_e}{K_m/p + S_e}$$

As $S_e$ is the measured substrate concentration the observed effective Michaelis constant ($K_v$)

$$K_v = K_m/p$$

Thus partitioning of the substrate towards the polymer will result in the observed value $K_v$ being lower than the intrinsic $K_m$ of the enzyme; partitioning away causing a higher $K_v$.

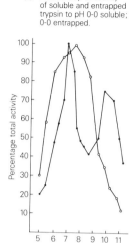

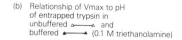

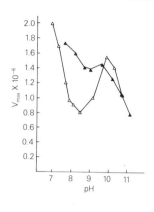

(a) Relationship of activity of soluble and entrapped trypsin to pH 0-0 soluble; 0-0 entrapped.

(b) Relationship of Vmax to pH of entrapped trypsin in unbuffered ▵——▵ and buffered ▲——▲ (0.1 M triethanolamine)

**Fig. 14.9** Relationship of activity to pH of soluble (○) and polyacrylamide gel entrapped (●) trypsin, (a) expressed as % total activity; (b) $V_{max}$ for the immobilized trypsin in the presence (▲) and absence (△) of 0.1 mol dm$^{-3}$ triethanolamine buffer. (From Trevan, M. D. and Groves, S. (1979))

If the partitioning is due to electrostatic interactions then the value of $p$ will tend towards unity as the ionic strength increases. Hence at high salt (or even high substrate) concentrations the difference between $K_v$ and $K_m$ will disappear. This is clearly shown in Fig. 14.11 which describes the effect of ionic strength on the observed value of $K_v$ for immobilized bromelain. Interestingly, the value of $K_v$ plateaus below the intrinsic $K_m$ and the investigators attributed this modification of the intrinsic $K_m$ as a direct result of immobilization. Where the substrate itself has a significant effect upon the value of $p$, with the result that the observed $K_v$ value tends towards the intrinsic $K_m$ as the substrate concentration is raised, double reciprocal plots of reaction velocity versus substrate concentration will be S-shaped rather than linear.

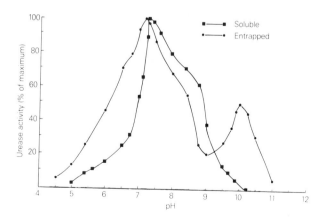

**Fig. 14.10** Relative activity of soluble and polyacrylamide gel entrapped urease. (From M. D. Trevan, unpublished data)

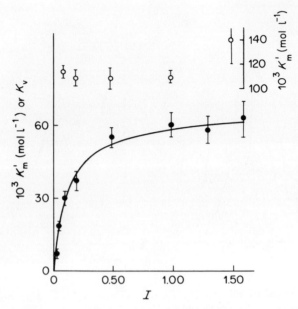

**Fig. 14.11**   Effect of ionic strength on apparent Michaelis constant $K_v$ of bromelain in dilute solution (○) and immobilized to carboxymethylcellulose (●), acting on benzoylarginine ethyl ester. (Reproduced with permission from Wharton, Crook and Brokelhurst, 1968)

*Diffusion limitation* of the substrate will similarly produce changes in the observed $K_v$, but in this case the value of $K_v$ will invariably be greater than the instrinsic $K_m$. Viewed simplistically this is the result of the fact that in order to diffuse into the enzyme's microenvironment the substrate must establish a concentration gradient such that its concentration in the bulk phase will be greater than that around the enzyme. Therefore, in order to attain the substrate concentration which gives half the maximum velocity for the immobilized enzyme ($K_v$) the substrate concentration in the bulk phase must exceed that which would give half the maximum velocity for the free enzyme ($K_m$). In essence, the observed reaction rate will be limited either by the rate of diffusion of the substrate to the enzyme or the intrinsic rate of the enzyme whichever is the lower. Which of these is lower will depend for a given system on two factors, the intrinsic level of activity of the immobilized enzyme and the substrate concentration in the bulk phase. Let us take the example of an enzyme immobilized at an impervious surface where substrate diffusion through the unstirred layer surrounding the enzyme is limited. The maximum attainable substrate diffusion velocity will increase linearly with an increase in the substrate concentration in the bulk phase ($S_e$), while the maximum attainable intrinsic reaction velocity increases with increased $S_e$ in a typical hyperbolic manner. Figure 14.12 shows both these trends plotted on the same graph. The effective reaction velocity tracks the lower of these two lines; this is controlled at values of $S_e$ below $A$ (see Fig. 14.12) by substrate diffusion while above $A$ by the intrinsic rate of reaction.

Several features now become apparent. First, the effect of substrate diffusion limitation may be overcome by raising the bulk phase substrate concentration or lowering the intrinsic enzyme activity. Second, double reciprocal plots of velocity versus substrate concentration will be curved. Third, increasing the temperature will

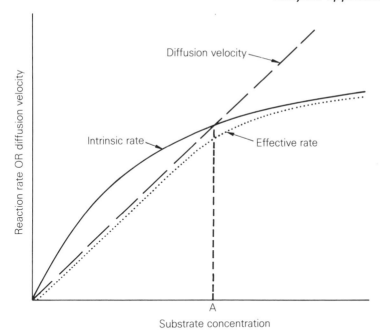

**Fig. 14.12** Relationship between intrinsic reaction rate (———), effective reaction rate ($\cdots\cdots$) and diffusion velocity ($----$) for an immobilized enzyme

enhance substrate diffusion limitation, because the intrinsic enzyme activity will increase by a greater proportion than will diffusion velocity. To double the intrinsic activity of the enzyme requires a mere 10 °C rise in temperature; double diffusion velocity would require the temperature to be raised by 300 °C (from 27 °C).

14.2.4.5 *Maximum velocity* ($V_{max}$). The agile-minded reader may by now have concluded that the maximum velocity of an immobilized enzyme ($V_s$) will be the same as that of the free enzyme ($V_{max}$) direct effects of immobilization on the enzyme apart. However, it must be noted that this will not always be the case. Two factors may give rise to $V_s$ differing from $V_{max}$, practical limitation on the solubility of the substrate and variation of the enzyme's stability (real or apparent) as a consequence of immobilization.

*Solubility of the substrate* may limit the maximum bulk phase substrate concentration, giving rise to the situation where the microenvironmental substrate concentration cannot be raised to the value required for maximal enzyme activity. In such cases $V_s$ will be lower than $V_{max}$.

*Enzyme stability* has its effect on the measured velocity of reaction because during the finite time over which the velocity is measured the enzyme may become denatured. This applies particularly at higher temperatures. Thus, if immobilization effectively stabilizes the enzyme, it may not be denatured over this time period and thus reaction velocities (and hence $V_s$) may appear raised (with respect to the free enzyme's $V_{max}$). This may be the explanation of the variation of $V_s$ values for immobilized amino-acylase as shown in Table 14.8, p. 227.

14.2.4.6 *Inhibitors and activators*. Here again partitioning or diffusion limitation will have their effect.

*Partitioning* of inhibitors or activators will enhance or diminish their effect depending on whether they are concentrated or expelled from the enzyme's microenvironment. Such partitioning will be affected by factors such as the ionic strength in exactly the same way as substrate partitioning described above.

*Diffusion limitation* of inhibitors or activators will, in general, only affect those molecules which also take part in the reaction—that is, substrates and products—because, eventually, homogeneous distribution will be attained. One notable exception to this is that molecules of high molecular weight may be excluded by gross steric exclusion. Thus an immobilized enzyme may show the same sensitivity to a low molecular weight inhibitor as its free form, but may be relatively unaffected by an inhibitor of high molecular weight.

Where an immobilized enzyme suffers from product inhibition then diffusion limitation, causing the accumulation of product in the enzyme's microenvironment, will exacerbate the inhibition. To add insult to injury many products are competitive inhibitors and if the product is subject to diffusion limitation so may the substrate. These two effects then work synergistically as the product concentration is highest where the substrate concentration is lowest. However, if the bulk phase product concentration is raised to a sufficiently high level the effect of diffusion limitation may be lost because the intrinsic enzyme activity is reduced to the extent where diffusion is no longer limiting.

It is actually possible to realize the situation where, if diffusion limitation of the substrate is severe, competitive inhibition, whether by product or substrate analogue, is never seen. This can occur because the bulk phase substrate concentration required

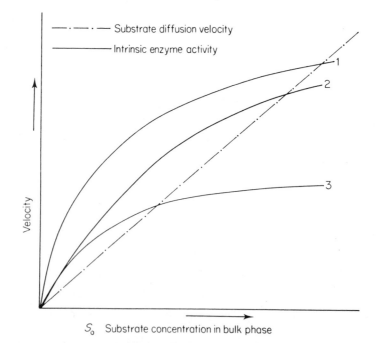

**Fig. 14.13**   Schematic representation of the interaction of substrate diffusion limitation and chemical inhibition for an immobilized enzyme. Solid lines represent the intrinsic enzyme activity of immobilized enzyme, uninhibited (1), subjected to competitive inhibition (2), and subjected to non-competitive inhibition (3). Clearly competitive inhibition may not be apparent if substrate diffusion limitation is severe.

(Reproduced with permission from Trevan, 1980)

to overcome the diffusion limitation also overcomes the effect of the inhibitor. This obviously will not be the case with a non-competitive inhibitor (Fig. 14.13). Thus the effect of substrate diffusion limitation and competitive inhibition may be antagonistic.

Substrate inhibition in the presence of substrate diffusion limitation can give rise to the observation of hysteresis loops in plots of substrate concentrate versus effective reaction velocity, because of the potential for multiple steady state rates for a given substrate concentration. It will suffice for present purposes to record that in general substrate inhibition, for an immobilized enzyme, will set in at higher bulk phase concentrations of substrate that for the free enzyme, and that this effect can find useful application in enzyme reactors. For a further discussion of this curious aspect of immobilized enzyme behaviour the reader is referred to Trevan (1980).

14.2.4.7 *The effect of immobilization on cells.* There is little that can be said about the effect of immobilization upon the behaviour of viable cells, because at this time this subject has not been systematically studied. However a few general points may be made.

The most common methods for viable cell immobilization involve some sort of entrapment within a hydrogel polymer. It is highly likely therefore that diffusion limitation will cause the microenvironment surrounding the cell to be quite different to the bulk phase in a number of ways.

Gas exchange between polymer microenvironment and bulk phase may be limited, particularly where the cells have a high metabolic activity. Thus $O_2$ and $CO_2$ levels may be substantially different in the cells microenvironment and this may directly

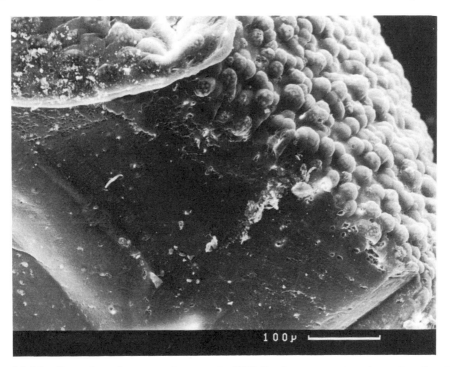

**Fig. 14.14** Scanning electron micrograph (SEM) of cross-section through a bead of calcium alginate immobilized algal cells, 4 days after immobilization; concentration of the cells at the periphery and their absence in the centre of the bead can be clearly seen. (Photo by C. W. Lilley)

affect their metabolism and metabolic output. For photosynthetic organisms light intensity may be reduced, not so much by the polymer but by self-shading of the cells deeper in the polymer by those on the outside. Metabolizing cells will often produce acid or alkali, and thus the pH of the microenvironment may differ from that of the bulk phase. This too may affect not only metabolic rate but also metabolite productivity. Hormones, extracellular regulators or enzymes will all tend to be accumulated around the immobilized cell with a variety of effects. For example, immobilized plant cells excreting endogenous hormones may thus provide themselves with a more beneficial environment when immobilized. An immobilized organism excreting an extracellular digestive enzyme may have that enzyme accumulate in its microenvironment thus enhancing uptake of an essential nutrient (which may anyway be present in a lowered concentration because of diffusional restrictions on its mobility into the polymer). The accumulation of secondary metabolites may switch off their own synthesis, and may result in immobilized cells being less productive than free cells. Conversely, limitation of growth because of nutrient limitation may switch cells into a stationary phase in which secondary metabolite production may be enhanced (see Section 6.3).

It is hardly surprising then that the potential limitation of nutrients or light, and the accumulation of toxic or inhibitory products in a cell's microenvironment as a result of immobilization, results in most of the immobilized cells in an old preparation being concentrated at the periphery (Fig. 14.14). This effect is probably due to cell death in the centre of the particle coupled to cell growth at the periphery rather than cell migration.

## 14.3  Soluble or immobilized?

If immobilization of biocatalysts is so difficult and empirical and so often results in unpredictable alterations in their behaviour, why bother? The reasons lie in the ability of immobilization to improve an enzyme's useful characteristics (see also Sections 14.1.2 and 14.1.3). The order of importance of these characteristics will depend in part upon the use to which the enzyme is to be put. At times and for certain applications, for example processing of high molecular weight polymers or gene manipulation, immobilized enzymes are not only inappropriate but also useless; for other applications, for example in biosensors, immobilization is mandatory; for most applications, however, whether or not immobilization should be used is a matter of cost and effectiveness. Before we consider the particular virtues of immobilized enzymes, we must begin by discussing the problems of soluble enzymes.

The application of soluble enzymes is limited by several factors: relative instability; cost; impracticability of recovery from the product thus dictating single use; inability to retain the enzyme within a stream of substrate, thus prohibiting continuous processing. Immobilization may overcome many, if not all, of these problems. Stability may be enhanced even if this is only in the operational sense rather than real inherent stabilization. This enhancement of stability is a prerequisite to the second advantage of immobilized enzymes, their ease of removal from the product and therefore their potential for reuse. Obviously it is no good having a reusable catalyst if it is not sufficiently stable to permit reuse. Reusability of an immobilized enzyme is one of the factors which offsets the potential high cost of the enzyme. It should be apparent that the more unit operations the enzyme can perform (for example, the more substrate it can process) the lower the proportional enzyme cost per unit product. The ability to retain, by immobilization, the enzyme within a moving stream of substrate makes continuous processes feasible with all their inherent economic advantages. The use of packed bed reactors, feasible only with immobilized enzymes, can lead to greater product yield and purity (see Section 14.6.1). Finally, and for the future perhaps the most important factor, there is the ability to produce by immobilization a change in an enzyme's behaviour. For example, the change in pH optimum of an enzyme as a consequence of a particular immobilization process may allow an enzyme, which normally operates around neutrality, to be used to process alkaline solutions. This may remove a need to alter the pH of the substrate solution or may be used as a device to help limit contaminating microbial growth. Substrate diffusion limitation is the key to biosensor linearity (see Section 14.7.2), or may be used to desensitize an enzyme to the fluctuations in pH which can occur in large-scale processes, thus maintaining the constancy of product yield without the need for precise and costly pH control. Partitioning of the substrate to the enzyme's microenvironment and/or product away from it will increase both product yield and substrate to product conversion by reducing the extent of product inhibition.

Many useful bioconversions involve substrates and products which are sparingly soluble in water. In order to improve conversion and yield organic solvents must be used as a carrier for such materials and thus in order to make a process economic the enzyme (or cell) must be stable in organic solvents. Immobilization may enhance this stability.

Most of the above considerations apply equally to immobilized cells, in particular the potential to maintain an advantageous microenvironment around the cell. In addition, plant and animal cells are slow growing and therefore the ability to improve the stability of the cells and allow their continued use is of vital economic importance. It is also possible when using immobilized viable cells, by allowing initially the cells to grow within the polymer matrix, to increase the cell mass per unit volume of reactor well beyond the practical limits of most free cell suspension cultures. This again will increase productivity.

## 14.4 Cells or enzymes?

Assuming that immobilization is considered desirable/cost-effective, the next question to arise is what form of biocatalyst to use, cells or enzymes? There are no hard and fast rules and the answer lies ultimately in consideration of overall costs. Throughout this discussion we are of course hampered by the conundrum 'when is an enzyme not an enzyme' to which the reply must be 'when its a cell'.

Biocatalysts come in a number of forms, the pure enzyme, crude cell extract, intact non-viable cell (usually microbial), viable microbial cell, viable plant cell, viable animal cell. Biocatalytic processes vary in complexity from simple single-step conversions, single-step conversions involving coenzymes, multistep processes for biotransformation of added precursors, *de novo* synthesis of compounds from simple carbon sources. The art is in the appropriate matching of biocatalyst type to process. Table 14.1 highlights some of the clues. It should be apparent from this table that the choice between cell or enzyme catalyst is a result of the interaction of a number of considerations: number of steps in the process; requirement for coenzymes; importance of contaminating reactions; cost; stability; catalytic efficiency. We can illustrate these interrelationships by selecting appropriate examples.

**Table 14.1**  Comparison of applications of various immobilized biocatalysts

| | *Enzymes* | *Microbial cells* | *Plant cells* | *Animal cells* |
|---|---|---|---|---|
| Immobilization method | Possible by wide range of techniques | Possible mostly by entrapment or adsorption | Entrapment in biogels; few other methods tested | Possible by adsorption |
| Inherent stability of un-immobilized form | Low | Medium → high | Medium | Low |
| Stabilization of bioca-talyst upon immobilization? | Yes, but apparent rather than real | Some stabilization | Real, marked stabilization | ? |
| Single-step processes | Yes, mostly hydrolytic reactions | Yes, mostly hydrolytic reactions with non-viable cell | Possible but enzymes or microbial cells probably more efficient | Not efficient |
| Single-step reactions involving (expensive) coenzymes | Possible, but still pro-hibitively costly | Possible, but fermenta-tion may be cheaper | Possible, may be uni-quely useful for process-ing waste plant products | May be possible but very costly |
| Multistep processes for biotransformation of added precursors | Perhaps technically feasible but prohibi-tively expensive | Possible, but traditional fermentation may be cheaper | Possible (see above); immobilization probably more efficient than fermentation | May be possible |
| *De novo* synthesis from simple carbon sources | No | Possible, but fermenta-tion almost certainly cheaper | Possible and probably cost effective if price + market volume is high enough | Essentially protein products |
| Major actual or poten-tial application | Probably confined to single-step mostly hydrolytic reactions | Used as 'unpurified' enzyme or possibly for single-step process in-volving coenzymes | Multistep process or, *de novo* synthesis | Production of desirable animal proteins |
| Type of product most likely to be produced | High volume organic compounds produced by single-step process | Medium to high volume organics possibly involv-ing single-step or multi-step reactions | Medium to low volume high cost plant products, photosynthetic processes | As above |
| Alternative production method | Chemical processes immobilized microbial cells | Fermentation | Agriculture, fermentation but not for all products | Extraction from animal tissue, genetic engineering |
| Long-term prospects | Good but limited | Advanced fermentation* may be more cost-effective | Immobilized cells may be more cost-effective | Genetic engineering may be ultimately more cost-effective |

* Except in processes involving organic solvent soluble products/substrates when immobilized microbial cells may be uniquely useful

If the process is a simple single-step catalysis, not involving coenzymes, then using a purified enzyme may have a number of advantages. With only one enzyme there will be no possibility of undesirable side-reactions occurring which may greatly simplify product purification. With pure enzyme high loadings of immobilized catalyst may be obtained. This may have a number of benefits, for example the use of smaller reactors or shorter process times. However, excessively high loadings may be wasteful as the immobilized enzyme may in practice be substrate diffusion-limited. This in itself may actually be advantageous as it may lead to an effective enhancement of operational stability, thus simplifying the process control and productivity. Set against these advantages is the cost of purification of the enzyme and the possibility that, for many membrane-bound enzymes, purification leads to reduced stability. In practice it may be cheaper to use a crude cell homogenate or whole non-viable cells, and remove contaminating enzymes by, for example, selective heat treatment (as in the removal of urocanase activity from *Achromobacter liquidum* cells containing histidine ammonia lyase activity) and to enhance the specific activity of the preparation by partial lysis of the cell membrane.

It is perhaps significant that most of the major large-scale uses of immobilized biocatalysts (such as glucose isomerase, penicillin acylase; see Section 14.7.4) employ immobilized non-viable microbial cells rather than purified enzymes.

It is when we turn to the more complex reaction system, perhaps requiring coenzymes that we find cells have their day. At the most complex level, for example, the production of alkaloids from simple carbon sources by plant cells, it is common sense to use the whole cell rather than to take the metabolic pathway apart and then have to reassemble it. However, it is also at the level of the one or two-step reaction system requiring coenzymes that immobilized cells have a particularly bright future. Although it is theoretically possible to retain and regenerate coenzymes within an enzyme reactor, why go to all the considerable bother and expense, when a cell is in essence a semipermeable bag containing enzyme, coenzymes and the necessary complementary recycling enzyme systems. Thus it is that whole cells of *Nocardia* spp. have been immobilized and employed as catalysts for the hydroxylation of steroids.

It would seem likely therefore that the future applications of immobilized enzymes as biocatalysts will be limited to simple, probably hydrolytic, reactions or to applications (such as biosensors) where high specificity of reaction is required. For more complex reactions immobilized 'cells' would appear to have a brighter future.

## 14.5  Stabilization

The first question we must ask is what is meant by the term stability? This may seem an obvious question but, to judge by the variety of implied definitions extant in the literature, it would appear that the term stability has a variety of subtly different meanings. In this context, however, we shall define stability as the ability of an enzyme to retain its catalytic activity; how one may enhance the stability of an enzyme is the topic of this section.

The next question is perhaps equally obvious. What factors can make an enzyme lose its activity? There are many: microbial digestion; hydrolysis by proteases (including autolysis); 'poisoning', for example by metal ions or oxidation reactions; aggregation and precipitation of the enzyme (for example, by organic solvents); and unfolding of the molecular structure of the protein. The first four of these factors may be fairly simply countered by making the enzyme inaccessible to the inactivating agent. For example, immobilization (see Section 14.2) will surround the enzyme with a cage that will physically prevent contact with micro-organisms and hydrolytic enzymes, or in the case of an immobilized proteolytic enzyme, prevent intermolecular contact and hence autolysis. Alternatively, the surface of the enzyme may be modified chemically, for example by alkylation or glycosylation, thus covering the possible sites of action of proteases on the enzyme's surface. Equally the polymer matrix of an immobilized enzyme may well, for steric or electrostatic reasons, prevent the poisoning of an enzyme by excluding or preferentially binding the poison, be it hydrogen or metal ions, organic molecules of even oxygen. Immobilization may also prevent inactivation caused by aggregation of the enzyme molecules brought about by, for example, organic solvents, either by physically holding the bound enzyme molecules apart or by negatively partitioning the solvent (that is, keeping it away from the enzyme).

When it comes to the prevention or elimination of the unfolding of an enzyme's molecular structure one is positively spoilt for choice of approaches. Before we consider these, however, a brief word about the unfolding (denaturation) of proteins. Denaturation can be brought about by a number of physical (temperature, pH, ionic strength) or chemical (urea, guanidine, organic solvents) means. In principle, however, however, their mode of action is similar: the disruption of the various electrostatic, hydrophobic, hydrogen bonds etc. which hold the enzyme molecule in its native shape, thus allowing the peptide chain to unfold and new associations to be made. This similarity of action of course means that a successful attempt to stabilize an enzyme against the action of, for example, urea, may stand a good chance of stabilizing it to increased temperatures. The trick is in effect to reinforce the native structure of the enzyme, and there are a number of ways in which this can, in theory at least, be achieved.

The precise mode of deactivation has been the subject of various hypotheses. At its simplest it is assumed that deactivation is a single step, first order process represented by

$$E \xrightarrow{k_1} E_1$$

Where $E$ is the enzyme's specific activity and in this case $E_1 = 0$; $k_1$ is the first order deactivation rate constant described in the equation

$$\frac{-dE}{dt} = k_1 E$$

Henley and Sadana (1985) have extended this approach to account for some of the less expected results of enzyme stability studies. They proposed a two-step model

$$E \xrightarrow{k_1} E_1 \xrightarrow{k_2} E_2 \quad \text{where} \quad \frac{E_1}{E} = \alpha_1, \quad \text{and} \quad \frac{E_2}{E} = \alpha_2$$

where either $\alpha_1$, or $\alpha_2$ could be greater or less than 1 and where $k_1$ or $k_2$ could vary independently $\geqslant 0$.

Assuming that the enzyme activity $(a)$ measured at any time

$$a = \frac{E + \alpha_1 E_1 + \alpha_2 E_2}{E_0}$$

where $E_0$ is the specific activity at time $= 0$,

by integration, the values of $E$, $E_1$, $E_2$ can be described in terms of $k_1$ and $k_2$ and any specific case can be modelled by substitution.

The result was to describe a variety of patterns of 'deactivation' where plots of log activity against time could show linear or complex loss of activity or even enhancement of activity (Fig. 14.15), which fitted experimental observations.

### 14.5.1 *Thermophiles*
One of the easiest ways of achieving something is to get someone else to do the job for you! Thus, the simplest way of obtaining a more stable enzyme preparation is to identify and isolate the enzyme from a thermophilic organism. Thermophiles characteristically contain enzymes of high thermal stability, which also often demonstrate superior stability to other denaturing agents. The reason for this

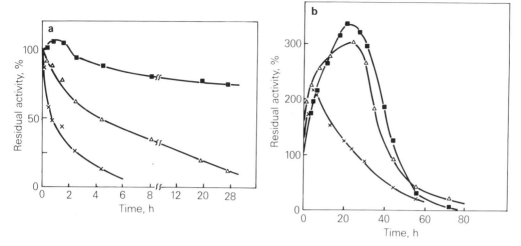

**Fig. 14.15** Effect of pH on the stability of (a), soluble and (b), immobilized 3-phosphoglycerate kinase at 25 °C and ■, pH 7.0; △, pH 8.0; ×, pH 9.0. Experiments were carried out in 0.1 mol dm$^{-3}$ triethanolamine/HCl buffer. Enzyme concentrations used were 20 $\mu$g cm$^{-3}$ for the soluble enzyme and 12 mg solid cm$^{-3}$ for the immobilized enzyme. (Reproduced with permission from Simon *et al.*, 1985)

increased stability is not, at present, well understood, but it seems unlikely that it is due to additional intramolecular covalent bonds (that is, disulphide bonding) as many thermostable enzymes contain fewer disulphide bonds than the same enzyme isolated from mesophiles. It could be suggested that thermostable enzymes are built to be intrinsically stable without the need for extra covalent bonding. That is, by virtue of adjustments to the enzyme's primary structure a greater number of non-covalent intramolecular bonds are formed, possibly bonds less susceptible to damage by heat. In this it seems likely that the enzyme relies more on internal hydrophobic interactions and less, for example, on hydrogen bonds.

### 14.5.2 *Enzyme engineering*
A related approach to the above is the technique of enzyme engineering (see also Sections 11.4 and 12.3.2). Here, however, the scientist gives nature a hand. In principle the approach is straightforward: isolate an enzyme from a variety of sources and investigate both the thermal stabilities and the structures of the enzyme; attempt to relate structural features to stability; predict what changes in amino acid composition are necessary to increase stability; isolate the gene for the enzyme and by the process of site directed mutagenesis alter the DNA appropriately; clone the gene back into an appropriate organism and investigate the stability of the resulting enzyme. In practice there are major obstacles to overcome. As yet little is known about the relationship between structure and stability, and it will probably be some time before such methods become routine (see Mozhaev and Martinek, 1984).

### 14.5.3 *Enzyme modification*
If effective modification of the enzyme's primary structure and consequent stabilization is at the moment more dream than reality, then direct chemical modification of the native enzyme as a means of achieving stabilization is a well-practised and

refined art. Martinek and Berezin (1977) have described the thermal stabilization of $\alpha$-chymotrypsin by alkylation, with acreolin, of free amino groups on the enzyme. The Schiff's base formed was stabilized by reduction with borohydride. A total of 15 amino groups per molecule could be alkylated. Each amino group had a different reactivity (presumably a result of their location in the enzyme) and, by altering the reaction conditions, any number between 1 and 15 of the groups could be alkylated. Alkylation of up to five amino groups had little effect upon stability; alkylation of between five and 13 resulted in a progressive, exponential increase in stability such that, with 13 groups alkylated the enzyme was 120 times more stable than the non-alkylated form. However, alkylation of all 15 amino groups resulted in the enzyme being no more stable than the native form. Martinek and Berezin (1977) in the same paper also describe an extension of this technique where not only are groups on the enzyme surface modified, but attempts are made to rigidify the molecular structure by intramolecular crosslinking with bifunctional reagents. The principle has been widely applied and is essentially similar to bolting girders or plates to the outside brickwork of a building to prevent it from collapsing. Many bifunctional reagents have been used; in essence a bifunctional reagent is a linear molecule with a reactive group at either end (for example, dialdehydes, *bis*diazonium salts, diioscyanates; Fig. 14.16). Much of the early work was empirical in nature, and in most often slight or negligible stabilization was achieved and was sometimes merely a result of single point modification of the enzyme. The secret of this form of stabilization, it was discovered, lies in having the correct size of bracket; different enzymes require different length bifunctional reagents, presumably because the distance between reactive groups on the surface of the enzyme varies according to the enzyme.

Torchillin *et al.* (1977) describe a system in which the stability of $\alpha$-chymotrypsin was enhanced 40-fold by reacting the carbodiimide-activated enzyme with diamino-butane $(NH_2(CH_2)_4NH_2)$, forming an amide bond between the diamino compound

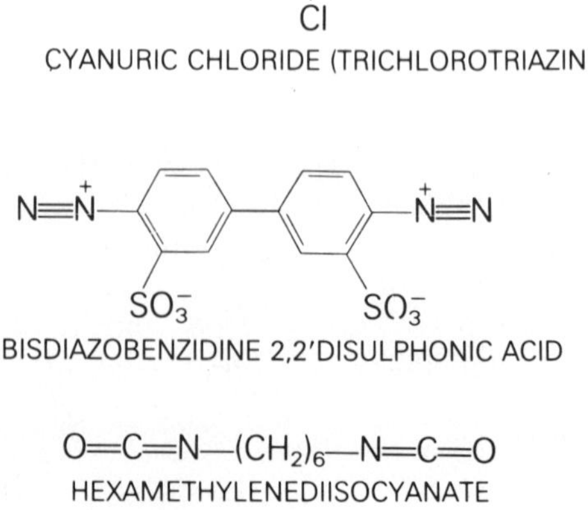

$$CHO \cdot (CH_2)_3 \cdot CHO$$
GLUTARALDEHYDE

CYANURIC CHLORIDE (TRICHLOROTRIAZINE)

BISDIAZOBENZIDINE 2,2'DISULPHONIC ACID

$$O{=}C{=}N{-}(CH_2)_6{-}N{=}C{=}O$$
HEXAMETHYLENEDIISOCYANATE

**Fig. 14.16**   Structures of some multifunctional reagents

**Table 14.2** Effect of alkyl chain length of diaminoalkyl cross-linking reagent on the stability of native and succinylated α-chymotrypsin; stability is inversely related to thermodeactivation rate constant (from Torchillin *et al.*, 1977)

| Chain length of cross-linking alkyldiamine | Rate constants of thermodeactivation $min^{-1}$ | |
| --- | --- | --- |
| | Native chymotrypsin | Succinylchymotrypsin |
| Non-crosslinked | 0.25 | 0.25 |
| 0 | 0.48 | 0.05 |
| 2 | 0.10 | 0.01 |
| 4 | 0.08 | 0.04 |
| 5 | 0.15 | 0.05 |
| 6 | 0.24 | 0.09 |
| 12 | 0.27 | 0.07 |

and carboxylic acid group on the enzyme's surface. Both the monoamine 1-aminopropan-3-ol and diamino-duodecane ($NH_2(CH_2)_{12}NH_2$) decreased the thermal stability of the enzyme, while diamino-pentane, diamino-hexane and diamino-ethane showed some stabilizing effect. When the experiment was repeated using succinylated α-chymotrypsin (succinic anhydride reacts with amino groups on the enzyme's surface thus increasing the density of available carboxylate groups) even greater stabilization was achieved, but this time the most effective crosslinking reagent was diamino-ethane (Table 14.2).

14.5.4 *Stabilizing agents*
There are, however, other ways of preventing unfolding of an enzyme's structure. Many examples exist which demonstrate that the presence of substrate or specific metal ions (for example, $Zn^{2+}$ or $Ca^{2+}$) may markedly stabilize the enzyme (see Section 13.2.8). Such effects may be explained by the additional bonds created by the binding to the enzyme of the added molecules. Such effects are obviously specific to a particular enzyme. It has also been demonstrated that stabilization of many enzymes may be effected by the addition of high molecular weight hydrophilic polymers such as dextrans or polyethylene glycols (see Schmidt, 1979). These latter agents presumably function by reducing the water activity around the enzyme, thus removing its effective hydration shell, and consequently restricting the enzyme's ability to alter its conformation. A fully denatured enzyme will have the maximum possible number of interactions between hydrophilic groups and water. If the water is made less available to the enzyme the 'denatured' conformation is less easily obtained and hence less likely to occur. In effect if the hydrophilic groups cannot interact with water because it has been removed, then they have no recourse but to interact with each other, thus maintaining the native conformation of the enzyme. Alternatively it may be a result of increased viscosity of the solvent. In this context it is interesting to note the observations of Zaks and Klibanov (1984) on the effect of water on the thermal stability of porcine pancreatic lipase. In an aqueous medium the lipase was instantly denatured upon heating to 100 °C, but when placed in a medium containing 2 M n-heptanol in trybutyrin and reducing the water content to 0.015%, the half-life of the lipase at 100 °C was more than 12 h! Progressively increasing the water content

to 0.3% resulted in a sharp decrease in this phenomenal thermal stability. However, their studies also illustrated a fundamental problem with many methods for enzyme stabilization, that by applying constraints to movements within the enzyme molecule in order to stabilize it, marked undesirable changes may be induced in its specific activity and specificity. Thus the 'dry' lipase showed a much less broad specificity than the 'wet' enzyme, and equally embarrassing the specific activity of the 'dry' enzyme at 100 °C was no better than the 'wet' enzyme at 20 °C. Thus this particular approach was irrelevant to the broader aim of increasing reaction rates by raising the reaction temperature (thus necessitating an increase in enzyme stability), but might be a useful way of enhancing the enzyme's specificity.

### 14.5.5   *Protein concentration*
One of the simplest ways of increasing a protein's stability is to increase its concentration. Forniani *et al.* (1969) studied the stability of glucose-6-phosphate dehydrogenase from different sources under a variety of conditions. The human enzyme at a concentration of 1 unit cm$^{-3}$ retained 90% of its activity for 90 days at 37 °C, but at 0.06 units cm$^{-3}$ had retained only 20% of its activity after 90 days. To put this in perspective the same enzyme was markedly stabilized by the presence of 0.5 mM NADP. The precise reason why more concentrated solutions of an enzyme (with the exception of proteases!) are more stable, or conversely why dilute solutions are relatively unstable, is not clear. It may have some relationship to the effect of soluble polymers described above, or it may be due to specific intermolecular interactions which will be more likely to occur at high enzyme concentrations. Whatever the reason this effect may well be, at least in part, responsible for the stabilization observed in the other forms of treatment of the enzyme described here, in particular the effects of immobilization on the enzyme (see below and Section 14.2.3).

### 14.5.6   *Immobilization*
There are two, related ways in which an enzyme can, potentially at least, be stabilized: intramolecular crosslinking (see above) and immobilization. Both of these methods supposedly operate by bracing the molecular structure, the former with internal ties and the latter with an external scaffolding. However, there are a number of traps into which the unwary investigator can fall, which may result in apparent or real stabilization of the enzyme molecule that may not be due to the crosslinking or immobilization *per se*. For example, any perceived stabilization might in fact be due simply to chemical modification of the enzyme's surface, to a reduction in autolysis and, in the case of the immobilized enzyme in particular, to an increase in the effective (local) protein concentration and/or the effects of substrate diffusion limitation (see Section 14.2.3). Furthermore, inactivation of an enzyme by heat, pH or denaturing agents can often not be described as a simple first order process (that is, a purely exponential loss of activity with time). It is not unusual for a 'pure' enzyme to exist in multiple forms, each of which would exhibit a different stability. The initial preparation of the crosslinked or immobilized enzyme may cause the total inactivation of the (predominant) labile form(s) of the enzyme, resulting in a preparation with an inherently more stable molecular structure than the 'free' form of the enzyme. For example, trypsin is usually purified as of mixture of $\alpha$ and $\beta$ forms, that differ in structure only by the degree of cleavage of the peptide chain. However, the $\beta$ form is some 100 times more stable than the $\alpha$ form. With these reservations in mind, it is not surprising, therefore, that there have been a multitude of claims of enzyme

stabilization as a result of immobilization. What is perhaps surprising is the unguarded optimism with which the enzyme technologist is sometimes seen to regard this method.

The principal considerations in achieving real enzyme stabilization as a result of immobilization seem to be those of multipoint attachment of the enzyme to polymer support and modification of the enzyme's microenvironment by the support.

The case for multipoint attachment appears, in retrospect, self-evident and is now well-documented. Clearly the more points of attachment there are between an enzyme molecule and its rigid polymer support the more movement will be constrained within the molecule. The relationships between multipoint attachment and structural rigidity of an enzyme have also been well-documented. However, there must be added three caveats to this principle. First, constraining conformational change in an enzyme may result in destabilization, if the enzyme is 'locked' in an unstable configuration; reports of destabilization of an enzyme as a result of immobilization are not uncommon. Second, most enzymes would appear to depend on some degree of conformational change for substrate binding and/or catalysis to occur. Restricting this ability for conformational change may well inactivate the enzyme, again a not uncommon event. Third, the surfaces of enzyme and polymer support matrix are unlikely to be complementary and thus, even if multipoint binding is achieved, only a small proportion of the enzyme's surface may be rigidified. Thus not much stabilization will be observed. Happily this latter problem is soluble. Martinek *et al.* (1977a and b) have described an ingenious method whereby the polymer support matrix is fitted to the shape of the enzyme molecule by forming the polymer around the enzyme. Two approaches were used. In the first $\alpha$-chymotrypsin was acrylolated, mixed with a solution of acrylamide monomers and polymerization initiated. The enzyme was thus incorporated into the structure of a polyacrylamide gel with a reported 200-fold increase in its stability when compared to the native and the acrylolated enzyme. When the same approach was tried with polymethacrylate, stability was enhanced 1000 times. The investigators also showed that varying the polyacryl gel concentration did not affect these stabilizations. In the second set of experiments unmodified $\alpha$-chymotrypsin was entrapped in polymethacrylate gels of various concentrations. Here the enzyme was presumed to be multiply bound to the gel by electrostatic interactions. The immobilized enzyme showed an enhanced stability that was dependent on gel concentration in a non-linear fashion. Thus gels of a concentration less than 20% showed little stabilizing effect. Between 20% and 50% there was a progressive stabilization. With a 50% polymethacrylate gel the theoretical half-life of the immobilized enzyme at 60 °C exceeded several hundred million years! In both these cases (but more particularly the latter) it is possible that some or all of the stabilization is due to a reduction in water activity in the microenvironment of the enzyme, in effect dehydrating the enzyme.

The microenvironment of the enzyme may also play a role in stabilizing the enzyme by preventing access to the enzyme of denaturing agents. Thus a polyanionic matrix may exclude hydrogen ions, consequently rendering the immobilized enzyme less susceptible to denaturation in acid solutions. Alternatively a polyelectrolyte matrix may act as a very effective local buffering agent, with a buffering capacity beyond the reach of most water-soluble buffers. In either case the effect will be to maintain an innocuous pH around the immobilized enzyme in spite of the extreme pH of the bulk phase solvent. Such an effect has been reported for lactate dehydrogenase immobilized to porous glass (Dixon *et al.* 1973). Similarly, stabilization against organic solvents by immobilizing an enzyme onto a hydrogel polymer (soluble or insoluble) has been proposed. However, successful studies are relatively few in number (see Douzou and Balny (1977) and Takahashi *et al.* (1984))). Stabilization is presumably due to exclusion of the organic solvent from the enzymes microenvironment and/or replacement of the enzyme's hydration shell by hydroxyl groups on the polymer matrix. Obviously it will be easier to prevent interaction between organic solvent and enzyme when the solvent is immiscible with water. Inactivation by oxygen is a more limited problem, but one which affects a group of enzymes of great potential use in the biological conversion of solar energy, the hydrogenases. Here too immobilization may be of value. Oxygen solubility in water is reduced if the ionic strength is increased, a process of salting out. Thus clostridial hydrogenase immobilized by adsorption onto polyethyleneimine cellulose had a half-life of 1 week when suspended in air saturated water compared to 4 min for the free enzyme (Klibanov, Nathan and Karmen, 1978). In effect the polyethyleneimine provides a microenvironment around the enzyme in which oxygen is virtually insoluble.

Finally, it is worth recording two curious observations on enzyme stabilization by immobilization. The first, reported by Cashion *et al.* (1982), is the gradual increase in specific activity of alkaline phosphatase as a consequence of immobilization on trityl-agarose, the final activity of the immobilized preparation was, after 12 days, 4.8 times that of the native enzyme. Carrea, Bavara and Pasta (1982), studying the stability of glyceraldehyde-3-phosphate dehydrogenase immobilized onto CNBr-activated Sepharose 4B, discovered that the activity increased by 2.3 times within an hour as a result of immobilization and was maintained at this level for over 30 h at 40 °C. Thus we can see that a large number of approaches may be taken to the problem of enzyme stabilization. The subject is at last leaving its empirical beginnings and is emerging as a conceptually rational subject area, albeit one fraught with artefactual traps for the innocent.

## 14.6   Biocatalyst reactors

In order to put a biocatalyst to good use one has to have somewhere to use it—the reactor. The design of appropriate reactors is an art in itself which depends on the form and nature of both biocatalyst and reaction. Obviously this subject has a lot in common with fermenter design, discussed elsewhere in this book (see Section 8.2). We shall therefore confine ourselves here to a consideration of the major practical differences between different reactors (principally for use with enzymes) and to the factors affecting choice of catalyst and choice of immobilization method where one is to be used.

### 14.6.1   *Reactor design*
Little needs to be said here about reactor design. All we shall do is to illustrate the range of possibilities, summarized in Fig. 14.17. Reactors may be subdivided into two categories, batch and continuous reactors.

14.6.1.1   *Batch reactors.* These are basically large stirred tanks into which are placed enzyme and substrate, the reaction is allowed to occur, the reactor drained and product and enzyme separated. In practice such reactors are used with cheap, soluble enzymes or where the substrate is in solid or colloidal form and thus immobilized enzymes are inappropriate. Despite the simplicity and low cost of such reactors they are not ideal for use in integrated continuous processes.

14.6.1.2   *Continuous flow reactors.* These are based on the principle of containment of

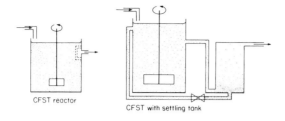

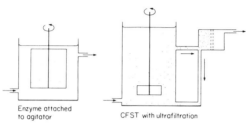

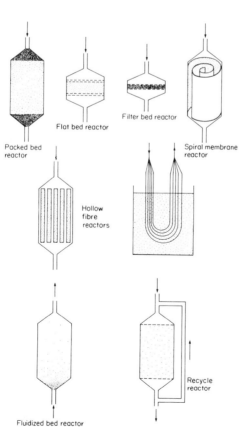

**Fig. 14.17** Types of continuous flow reactors. (Reproduced with permission from Trevan, 1980)

the biocatalyst in the reactor with the continuous addition of substrate and removal of product. Such reactors, which come in a variety of forms, can be integrated into the whole process which may then be amenable to automation. Continuous flow reactors are not, however, the ultimate in reactor design and may be quite inappropriate if the rest of the process is not continuous.

The catalyst can be held in the reactor either by immobilizing it in some way or by using the soluble form in conjunction with an ultrafiltration unit at the outlet point. By the judicious choice of substrate concentration, flow rate, biocatalyst activity, the productivity and percentage conversion of the reactor may be optimally adjusted. The main types of reactor are described and their peculiar advantages outlined below (see also Table 14.3).

14.6.1.3   *Continuous flow stirred tank reactor.* The most straightforward continuous reactor is the continuous flow stirred tank reactor. The catalyst is suspended in a large tank through which substrate flows, and is retained within the reactor by filtration, subsequent sedimentation, magnetic forces or by being attached to the stirrer paddles. Such reactors are easy to build and control and allow relatively easy replacement of spent catalyst. Good mixing can be obtained and thus diffusional limitations can be reduced to a minimum. Given a suitably immobilized enzyme colloidal or solid

**Table 14.3**   Comparison of reactor types

|  | CSTR | Packed bed reactor | Fluidized bed reactor | Hollow fibre reactor |
|---|---|---|---|---|
| Ease of control | Good | Moderate | Good | Moderate |
| Percent conversion of substrate | <93% | >98% | >98% | >98% |
| Yield with substrate inhibition | High | Low | Moderate | Low |
| Yield with product inhibition | Low | High | Moderate | High |
| Plugging | None | Possible | None | None |
| Use with colloidal substrate | Yes | No | Yes | Low molecular weight substrates only |
| Replacement of enzyme | Easy | Difficult | Difficult | Easy |
| Enzyme properties | Non-brittle | Incompressible | Non-brittle | Soluble |
| Void volume | High | Low | Moderate | — |
| Flow rates | Any | Any | High | High |
| Reactor stability | Good | Good | Poor | Poor |
| Mass transfer | Moderate | Poor | Good | Poor |
| Deviation from ideal characteristics | No | Yes | Yes | Yes |
| Pressure drop | None | High | Low | Low |
| Power requirement | Moderate | Low | High | Low |
| Scale up | Easy | Moderate | Difficult | Moderate |
| Automation | Moderate | Easy | Moderate | Easy |
| Ease of commission | Easy | Difficult | Difficult | Easy |
| Cost | Moderate | High | High | High |

substrates may be used, or alternatively it may not be necessary to remove solids etc. from the substrate stream. This form of stirred tank reactor does require a non-brittle immobilized catalyst to prevent attrition by the stirrer blades. In practice the void volume of tank reactors is of the order of 98% (that is, the catalyst occupies only 1 or 2% of the volume) which means that compared to packed bed reactors of the same productivity, continuous stirred tank reactors may have to be an order of magnitude larger (in order merely to incorporate sufficient catalyst). Not only is overall productivity low, but percentage conversion of substrate to product tends to be low. In a tank reactor the substrate is homogeneously distributed throughout the reactor at the concentration in the product stream (Fig. 14.18). Thus substrate concentration is uniformly low, and can never be reduced to near zero. Conversely product concentration is uniformly high and the presence of product inhibition of the catalyst may reduce productivity still further. In practice it is probably uneconomic to aim for conversion factors of more than 90% with this type of reactor.

14.6.1.4 *Packed bed reactors*. These come in a variety of designs but are all characterized by small size, high productivity even in the presence of product inhibition, low void volume, ease of automation. However, they do suffer from a number of disadvantages. Lack of easy access can make catalyst replacement awkward and environmental (particularly pH) control difficult. Fabrication and commissioning costs are high although running costs may be low. Particulate, colloidal or high viscosity substrate streams tend to block packed bed reactors, and in addition unless the catalyst is incompressible, channelling or blocking of the flow through the catalyst bed may occur. This latter problem and that of potential gas liberation (for example, by photosynthetic immobilized plant cells) can be overcome by using a sheet immobilized enzyme rolled up into a spiral and inserted longitudinally into the reactor, or by using skeins of immobilized enzyme fibres. Any mixing is dependent upon the flow rate and thus such reactors are susceptible to diffusional limitations as a result of poor mixing. Increasing flow rate to overcome this problem will reduce productivity and conversion unless either the reactor length is increased or some recycling of the stream is introduced. This last 'solution' may cause yet more problems

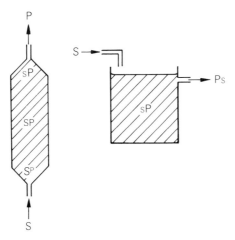

**Fig. 14.18** Substrate and product concentrations in stirred tank and packed bed reactors

as recycling effectively destroys that heterogeneity of product concentration which allows high productivity and conversions!

14.6.1.5  *Fluidized bed reactors*. These are in theory at least blessed with all the advantages of the packed bed reactor but none of the problems. These reactors work by the application of a high pressure substrate stream into the base of the reactor which lifts and mixes the catalyst bed. Thus channelling cannot occur; colloidal or even solid substrates could be used; mixing is improved and diffusion limitations reduced with the result that high biocatalyst loadings may be used efficiently and productivity of the reactor increased. Why then does not everybody use this form of reactor? The answer is: uncertainty and cost. Power requirements for fluidizing the catalyst bed are high, and although fluidized bed reactors work well on a laboratory scale it is difficult to predict the success of a scaled-up process.

14.6.1.6  *Hollow fibre reactors*. These reactors, based upon commercially available ultrafiltration units, where the catalyst is separated from the substrate stream by the semipermeable wall of the ultrafiltration hollow fibre, have recently begun to receive increased attention. Ultimately, these reactors may prove economically attractive because they can be used with soluble enzymes (which can be easily replaced), and may be of unique value in reactions involving coenzymes (see Section 15.1).

### 14.6.2  *Catalyst design*

The choice of catalyst for use in a reactor involves a number of considerations: the nature of the biocatalyst; the need for immobilization; the reaction; immobilization polymer support material; and, most critically, cost.

While it is impossible to generalize about biocatalyst design some useful observations can be made. In the end empiricism usually rules. The starting point is usually a knowledge of the reaction to be carried out, which in turn will dictate the type of catalyst to be used and whether this will be immobilized (see Sections 14.3 and 14.4).

14.6.2.1  *The enzyme*. This may be cheap and readily available and it may not be worth immobilizing it, being less costly to use it in the soluble form as an expendible reagent, as in the liquefaction of starch by amyloglucosidase. Immobilization should only be considered when the enzyme is costly and/or of limited availability. The question which arises, of course, is how costly is an enzyme? Two factors make up the expense of an enzyme: purity, which determines absolute cost, and the proportional cost of the enzyme as part of the total process cost. Thus an enzyme which may be expensive in absolute terms may contribute less than 0.5% to the total process cost and thus could be regarded as expendable, whereas a cheap enzyme, if it forms for example more than 10% of the total soluble batch process cost may be worth immobilizing and thus reusing. In addition the absolute cost of an enzyme usually rises with increasing purity, but so also will its specific activity. A highly active enzyme preparation will be able to process more substrate in a given time period, thus for a fixed yield of product the substrate stream needs to be in contact with the enzyme for a relatively short period of time (known as the reactor residence time). As this residence time is equal to the ratio of the void (or liquid) volume of the reactor to the flow rate, low residence times will allow either high flow rates (an advantage in packed bed reactors) or small reactor size (thus reducing costs).

Thus a balance must be struck between the cost of purification of the enzyme and the cost reductions available from the use of enzyme preparations of high specific activity. At this time actual industrial practice seems to favour the use of impure enzymes or whole cells.

14.6.2.2 *The polymer matrix.* The polymer matrix for immobilization should be inexpensive, preferably food grade material, stable, incompressible, non-brittle, retainable within the reactor, show no leaching, resist microbial attack, have the correct hydrophobic/hydrophilic characteristics, have a high surface area:volume ratio, bind large quantities of enzyme with no time-dependent loss, be of the correct physical form or shape, allow easy coupling of any enzyme or cell using non-toxic reagents! Not surprisingly some compromise is usually required; many of the listed characteristics are mutually exclusive. Thus mechanical stability is greatest in compressible polymers; surface area:volume ratio increases with decreasing diameter of particles, but so too does fragility. The majority of polymers used have a density close to one, and are therefore relatively difficult to retain in stirred tank or fluidized bed reactors. A great deal of effort and ingenuity has gone into the search for a high density immobilization polymer, the usual solution being to coat a high density particulate material (for example, iron filings) with a hydrogel shell.

The marriage of enzyme and polymer must then take account of practical limitations of what is possible and yet result in the closest to the ideal within the predetermined budget.

## 14.7 The application of biocatalysts

The applications of enzymes and cells can be divided into four categories: therapeutic uses, analytical uses, manipulative uses and their use as industrial catalysts. A comprehensive review of all of these aspects is beyond the scope of this text, so we shall settle for a selective description of some of the more exciting examples.

14.7.1 *Therapeutic uses*
Enzymes may be used in one of two ways in therapeutic applications, to replace an enzyme which is missing either as a result of some inborn error of metabolism or organ failure (enzyme replacement), or to remove some unwanted material from the body in the treatment of a disease (enzyme therapy).
14.7.1.1 *Enzyme replacement.* There are a multitude of diseases in which the absence of a specific enzyme causes the accumulation to toxic levels of endogenously occurring materials, or a malfunction of a metabolic pathway or metabolite transfer. There are a number of examples. The disease favism, a frequently fatal reaction to broad beans (*Vicia fava*), is caused by the lack of the enzyme glucose-6-phosphate dehydrogenase in the red blood cell. This impairs the cell's ability to regenerate glutathione and leads to cell membrane fragility and haemolytic anaemia. Tay–Sachs disease involves the absence of a specific hexoseaminidase, normally found in the lysozomes of the ganglion cells of the brain, whose function is to remove the carbohydrate residues of the ganglioside lipids as the first step in their breakdown. The ganglioside (specifically $GM_2$) levels in these cells slowly increases, disturbing brain function and usually resulting in the death of the individual before the age of 2 years. The synthesis of a range of steroid hormones from a common precursor, cholesterol, is a prime example of a branching metabolic pathway, where disturbance of the function of one of the branches can have dire consequences on the synthesis of the products of the other branches. Thus the full or partial loss in the adrenal cortex of an hydroxylase enzyme responsible for introducing a hydroxyl group onto the 21-carbon of the steroid ring structure leads to the condition known as virilization. Interestingly the missing enzyme is responsible for the synthesis of the glucocorticoids and mineralocorticoids

(such as aldosterone), but in addition to a loss of some of these hormones there is an excess production of the male sex hormones, leading to masculinization of female children and precocious sexual activity in males (usually about age 5–7 years). In addition to these genetically inherited diseases important enzyme activity may be lost in cases of tissue failure.

In theory at least such diseases could be treated by administering the appropriate enzyme. However, in practice, this approach is usually ineffective and can do more harm than good: there may be an allergic response to the exogenous enzyme; the enzyme may be rapidly cleared from the blood-stream; or it may be unable to reach or enter the appropriate target cell. Thus more subtle approaches must be made. In the case of virilization mentioned above, treatment is usually effected, non-enzymically, by the administration of aldosterone which both replaces the missing hormone and re-exerts the normal control mechanisms on steroid synthesis, thus preventing overproduction of the sex hormones.

In other cases, however, where the enzyme activity must be replaced, two approaches can be used. The enzyme must enter the body in a form where it is non-allergenic, retained within the body and, if an intracellular metabolite is to be attacked, directed to and absorbed by the appropriate cell. The enzyme can be incorporated into an extracorporeal shunt. In either case immobilization of the enzyme by entrapment or encapsulation may prove useful. Thus many attempts have been made to immobilize enzymes in liposomes (concentric spheres of phospholipid bilayers) which are targeted into specific cells by coating them in monoclonal antibodies specific to those cells. These liposomes are then injected into the circulation, they bind to and then fuse with the targeted cell membrane, and by the process of endocytosis the enzyme is released into cell. So much for the theory—in practice success is variable. However, in many cases such targeting is not necessary because the accumulated metabolite is found in the extracellular fluid, and thus either stable injected enzymes or an extracorporeal shunt may be used.

Whereas most such replacements will involve simple hydrolase or oxidase enzymes, some conditions involve coenzyme requiring enzymes. In these cases coenzyme regenerating systems must be incorporated with the enzyme (see Section 15.1). For example, the inherited deficiency of galactokinase has been treated experimentally by injecting coencapsulated galactokinase with ATP/ADP and pyruvatekinase as an ATP regenerating system.

It should be noted that not all enzyme deficiency diseases require enzyme replacement. Thus the lack, in most adults, of intestinal $\beta$-galactosidase, which leads to digestive disturbances due to the presence of undigested lactose, or phenylketonuria (PKU) caused by the inability to metabolize fully phenylalanine are most simply 'treated' by removing the offending substance from the diet.

14.7.1.2   *Enzyme therapy.*   Enzyme therapy involves adding an enzyme to the body which is either not normally present, or is not pathologically diminished, in order to remove material which may be causing or enhancing a diseased state. Thus treatment of certain leukaemias, in which the leukaemic cells is dependent for growth on exogenous asparagine, has been effected by the administration of bacterial asparaginase. Intravenous urokinase is used to stimulate a cascade mechanism which produces active plasmin (a protease which digests fibrin) in patients at risk from pulmonary embolism.

The field of enzyme application in medicine will yield great prizes in the coming years; already much has been achieved (the world market in 1981 for urokinase was £50 000 000, just half that for insulin).

14.7.2 *Analytical uses*
The analytical uses of enzymes are legion, and thus we shall once again concentrate on the novel or unusual aspects of this area. Traditionally enzymes have been used to catalyse the specific conversion of a compound of interest into one which can be more conveniently measured. The high stereospecificity of most enzymes allows the conversion of just one substance in a complex mixture of chemical similar substances, thus eliminating the need for costly separation procedures prior to analysis. Two forms of enzyme analysis are possible: end-point, and kinetic analysis. Briefly, end-point analysis involves the total conversion of substrate to product in the presence of sufficient enzyme to cause the reaction to be completed in a few minutes. This form of analysis is uncritical of enzyme concentration or the presence of inhibitors or activators. Kinetic analysis utilizes the relationship between substrate concentration and rate, and may be used to measure the enzyme's substrate concentration or reversible inhibitors of the enzyme (including inorganic molecules). For a full discussion of these aspects of enzyme analysis the reader is referred to Palmer (1981).

14.7.2.1 *Enzyme immunoassays*. Enzymes may also be used in the analysis of low concentrations of biological molecules such as hormones by coupling their catalytic ability to produce a measurable change in the analytical system with the specific binding ability of antibodies (immunoglobulins), the so-called enzyme immunoassays, ELISA (enzyme-linked immunosorbent assay) and EMIT (enzyme-multiplied immunoassay technique).

ELISA resembles radioimmunoassay in the sense that the enzyme activity is used as a convenient label tagged onto the antigen or usually antibody molecule. Antibodies in serum can be detected by first coating the inside of a plastic tube (or well on a plate) with appropriate antigen and adding to this the serum. Any antibodies present stick to the antigen. To this is then added enzyme-labelled antiglobulin which binds stoichiometrically to the antibody–antigen complex. Substrates for the enzyme are added and the rate of release of product is a measure of the amount of antibody present. Two variations of approach are possible to measure the level of an antigen. The first involves the use of a plastic tube coated with a known quantity of antigen, to which is added the sample antigen and a known quantity of enzyme-labelled antibody. The immobilized and sample antigens compete for the antibody such that the amount of antibody (and thus enzyme) that ends up bound to the tube is inversely proportional to the quantity of sample antigen. Alternatively, antibody is coated onto the inside of the tube, the sample antigen added which adheres to the antibody. Enzyme-labelled specific antibody is then added and binds stoichiometrically to the bound antigen. All of these techniques involve the production of class or species-specific enzyme-labelled antiglobulins or enzyme-labelled specific antibodies, a time-consuming and costly business. It has thus been proposed that the enzyme should be linked to protein A (produced by *Staphylococcus aureus*), a protein which binds specifically to subclasses 1, 2 and 4 of immunoglobulin G. This enzyme–protein A reagent could then be used to bind to any bound immunoglobulin in place of enzyme-labelled immunoglobulins. Research has even progressed to the point of trying to engineer genetically an enzyme-protein hybrid molecule to alleviate the necessity of combining the two chemically.

Enzymes used in this way include peroxidase, alkaline phosphatase, $\beta$-galactosidase and $\beta$-lactamase. The latter is particularly useful in the analysis of human samples as it is not normally present in them and in addition has a number of artificial substrates which are converted into highly chromogenic products.

14.7.2.2   *Autoanalysis*. Routine analysis of multiple samples in clinical biochemistry laboratories is usually carried out on an autoanalyser. Aliquots of sample and reagents are mixed and pumped automatically into the flow cell of the analyser (usually a spectrophotometer). However, despite its convenience such a process suffers from a major drawback—the cost of the enzyme. Not only is the enzyme lost with each sample, but in order to assure complete reaction within a sensible time span excess quantities of enzyme must be added. This problem can be alleviated by placing a coil of immobilized enzyme in the sample stream. To this end a number of companies produce enzyme coils (usually the enzyme is coupled onto the inside wall of several meters of thin nylon tubing), the first of which was a glucose dehydrogenase coil for monitoring blood glucose. In theory at least, several different coils could be used in series with repeated passes of the sample through the analyser flow cell(s) in order to analyse more than one substrate in the sample.

14.7.2.3   *Biosensors*. The latest hot favourite from the enzyme assay stables is the biosensor, the combination of biologically active material displaying characteristic specificity with chemical or electronic sensor to convert a biological compound into an electrical signal. Biosensors have been constructed to measure almost anything from blood glucose levels to the freshness of fish. We shall have to content ourselves here with a brief description of the principles of the modes of action illustrated by a few, hopefully well-chosen, examples. For a more detailed consideration the reader is referred to Carr and Bowers (1980), Guilbault (1980), Aston and Turner (1984), and Karube (1984).

The principle of the biosensor is one of those aesthetically beautiful concepts that are fiendishly awkward to apply in practice. Thus while the origin of the idea is usually attributed to Clark and Lyons in 1962, even today commercially available biosensors are thin on the ground. The object is to immobilize a biocatalyst in the direct vicinity of a sensor that can measure a reduction in one of the substrates or an increase in one of the products of the 'reaction' catalysed by the biosensor. The monitored substrate might be oxygen, the product ammonium ions, hydrogen ions, heat, light or even directly electrons. A variation on this theme is the immobilization of antibodies or enzymes onto the gate of a field-effect transistor which measures the change in charge resulting from binding of the antigen or substrate. The biocatalyst could be a purified enzyme, a whole cell, an organelle or any combination of these. The sensor can be anything, a simple carbon electrode, an ion-sensitive electrode, oxygen electrode, a photo cell or a thermistor. A multitude of such biosensors have been constructed from the fairly obvious, for example, the glucose electrode based on a combination of glucose oxidase and an oxygen electrode or the cholesterol electrode based on *Nocardia erythropolis* cells and an oxygen electrode, to the downright extraordinary, for example, the adenosine electrode based on a combination of mouse small intestine cells (containing adenosine deaminase) immobilized around an ammonia-sensitive electrode. All such electrodes have a number of similar features well illustrated by the glucose electrode.

A simple glucose electrode (Fig. 14.19), can be constructed by immobilizing a layer of glucose oxidase in polyacrylamide gel around a platinum oxygen electrode. When a solution of glucose is brought into contact with the electrode, glucose (and oxygen) diffuse into the enzyme layer and are converted to gluconolactone and hydrogen peroxide, lowering the oxygen concentration in the gel layer around the electrode. The rate of reaction, recorded as the rate of diminution of the oxygen concentration read by the electrode, is proportional to the glucose concentration in the sample. Such a device responds linearly to glucose concentrations over a range $10^{-1}$–$10^{-5}$ mol dm$^{-3}$ with a

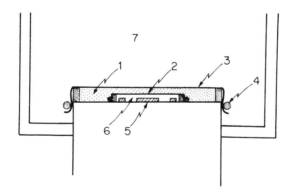

**Fig. 14.19** Diagram of a glucose-sensitive enzyme electrode: (1) immobilized glucose oxidase; (2) Teflon membrane; (3) cellulose acetate membrane; (4) 'o'-ring seal; (5) platinum oxygen electrode; (6) saturated KCl solution; (7) sample solution. (Reproduced with permission from Trevan, 1980)

typical response time of 1 min, and is stable for up to 4 months. Linearity of response is a result of the physical construction of the electrode; high enzyme loadings and diffusion limitation of the glucose mean that the rate of reaction is effectively controlled by the rate of diffusion of glucose into the gel, which is itself directly proportional to the glucose concentration in the sample. Constructing the electrode in this way will also enhance its operational stability (see Section 14.2.3). A few examples of biosensors constructed in this way are given in Table 14.4.

Such electrodes tend to be large and relatively clumsy devices and thus much effort in recent years has gone into the development of biosensors where a redox reaction catalysed by an enzyme is directly coupled to an electrode. The enzyme is presented with the oxidizable substrate only, and the electrons are transferred from the substrate to the enzyme and then directly to the electrode, rather than to the second substrate which is reoxidized by the electrode (Fig. 14.20).

**Table 14.4**  Typical enzyme biosensors

| Substrate | Enzyme | Stability | Response time | Range |
|---|---|---|---|---|
| Alcohol | Alcohol oxidase | 120 days | 30 s | $5-10^3$ mg ml$^{-1}$ |
| Cholesterol | Cholesterol oxidase/ cholesterol esterase | 30 days | 2 min | $10^{-2}-3 \times 10^{-5}$ mol dm$^{-3}$ |
| Uric acid | Uricase | 120 days | 30 min | $5 \times 10^{-3}-5 \times 10^{-5}$ mol dm$^{-3}$ |
| Sucrose | Invertase | 14 days | 6 min | $10^{-2}-2 \times 10^{-3}$ mol dm$^{-3}$ |
| Glucose | Glucose oxidase | 50–100 days | 10 s | $2 \times 10^{-3}-3 \times 10^{-6}$ mol dm$^{-3}$ |

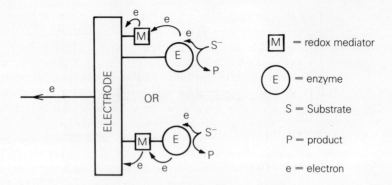

**Fig. 14.20**   Direct enzyme electrode

### 14.7.3   *Manipulative uses*

The specific catalytic abilities of enzymes have been employed for many years to manipulate other biological materials. On a large scale, such as the lysis of bacteria with lysozyme, this verges on the industrial use. However, on a small laboratory scale, much use is made of enzymes in experimental methodology. It would be impossible to summarize all those uses here, but the reader is referred to Section 11.2 for a consideration of the importance of enzymes in genetic engineering.

### 14.7.4   *Industrial uses*

In a book such as this which deals mainly with the biological principles underpinning biotechnology, it is not possible to give a detailed account of all the possible applications of enzymes. For those readers who require a detailed insight there are several texts available (Godfrey and Reichert, 1983; Poulsen, 1984). However some comment is appropriate.

Table 14.5 lists some of the more important and unusual applications of soluble enzymes as industrial catalysts, and little more need be said about these, except to point out that the majority of the enzymes used are hydrolases, mostly for food processing, and that some 25% of all enzymes manufactured end up in washing powders. This rather prosaic fact neatly illustrates the simple nature of much industrial enzyme technology, compared to the potential complexity which could be involved when (or if) more sophisticated enzyme based processes (for example, the synthesis of steroids) are introduced. The cost and instability of enzymes and coenzymes for such processes must almost inevitably mean the use of immobilized enzyme/coenzyme systems, so it is to these that we shall now turn.

Table 14.6 summarizes most of the present uses of immobilized enzymes in industrial processes. The first thought that might strike the astute reader is why are there so few present applications? The answer is straightforward: economics. Despite the undoubted technological advantages of immobilized enzymes and the wealth of research into immobilized enzyme based processes, the cost of actually setting up a process may be prohibitive, particularly if it is to compete with an existing, well-developed process. It is interesting to note therefore, that the largest immobilized enzyme process, the production of high fructose corn syrup using immobilized glucose isomerase, is one for which there was no competing technology. The Cetus Corporation subsequently patented another enzyme-based process to perform the same conversion, but this has failed to establish itself, largely because it had to compete with the existing glucose isomerase technology. In the event, the commercial success of the glucose isomerase process was made possible by the then high cost of the only other commercially available sweetening syrup, sucrose. By the time, several years later, that world sugar prices fell, high fructose corn syrup had established itself as a viable product with its own share of the 'industrial' sweetener market (about 50% in the United States). When the same process was first introduced into Europe, the European Commission placed such a high tariff on fructose syrups, to protect the sugar beet farmers, that production became uneconomic overnight!

Perhaps the most significant process listed in Table 14.6 is the production of acrylamide using immobilized nitrilase, for here we begin to move enzyme technology into new and largely uncharted territory; the use of enzymes to produce or manipulate materials more usually regarded as chemical than biological. This is the first real example of enzyme technology competing directly with traditional chemistry, despite the dubious present economics of the process, and is surely the way in which enzyme technology will develop over the next few decades. Examination of the rest of Table 14.6 reveals few other suprises apart from the dubious economics of many of the processes. Quite apart from the purely political considerations mentioned above, whether or not a process is economic will depend on a number of often interrelated factors, including raw material costs, product market price and volume, regional labour costs, transport costs, investment costs and the cost of purchasing or developing the relevant technology. For example, it may be technically feasible to develop a more economic process for the production of jasmine oil (which is almost literally worth its weight in gold). However, the cost of developing and setting up the process might well exceed the limited return possible, because of the low market volume for jasmine oil (see Table 1.2, p. 10). Labour and transport costs are self-explanatory except to add that, when dealing with a process converting a high volume low cost material into a low volume high cost product, transportation is cheaper with the production plant near the source of raw material rather than the market place. Capital investment costs are also a major factor; with borrowed money, a change of $\frac{1}{2}$% in interest rates can make a significant dent in projected profitability. Uncertainty and fluctuations in commodity markets can make the costing of raw materials and the value of products uncertain. Merely introducing the new technology changes the market place for, by creating a new demand for a raw material, the price is likely to rise and in producing more of a product into the market, is liable to force the price down. Thus at times of economic recession, investing in a new and expensive technology, whose profit is largely made by processing raw materials to products of higher value, involves a considerable act of faith!

Economics are also affected by the feasibility of process design, and one of the key parameters here is the operational life of the immobilized enzyme catalyst. Clearly a longer operational life reduces the necessity for frequent changes of catalyst, thus catalyst cost is reduced, as is the time spent changing the catalyst when the production facilities will remain idle. In practice this latter problem is overcome by running several reactors in parallel, staggering the time at which the individual reactors are brought on stream. Thus, in an array of, for example, eight reactors, seven may be operational at any one time whilst the eighth is off stream for catalyst regeneration. An additional advantage of such a reactor configuration is that the total yield and productivity of the system will vary less than with one large reactor, making the operation of upstream and downstream processes somewhat simpler.

It is appropriate to conclude these deliberations with a brief look at one well-established process, the formation of L-amino acids from DL-acylamino acids, by

**Table 14.5** Industrial uses of soluble enzymes

| Process or product | Amylase | Amylo-glucosidase | Catalase | Cellulase | β-glucanase | Glucose oxidase | Invertase | Lactase | Lipase | Pectinase | Protease |
|---|---|---|---|---|---|---|---|---|---|---|---|
| | | | | | | **Enzyme** | | | | | |
| Baking | Acceleration of fermentation; improved loaf volume, texture and structure | | | Extraction | | | | | | | Improved texture and loaf volume |
| Biomethanation | Pretreatment of wastes | Pretreatment of wastes | | Pretreatment of wastes | | | | | Pretreatment of wastes | Pretreatment of wastes | Pretreatment of wastes |
| Beer | Mashing | Low calorie drinks | | Extraction | Improved filtration | | | | | | Chill proofing, nitrogen control |
| Cereals | Precooked foods | | | | | | | | | | Condiment production |
| Cocoa/tea/coffee | Cocoa syrups | | | Cellulose hydrolysis in coffee drying, tea fermenting | | | | | | Fermentation, coffee concentrates | |
| Confectionery | Sugar recovery from scrap | | | | | | Fondant production | | | Fondant production | |
| Corn syrup | High maltose syrups | Production of glucose syrups | | | | | | | | | |
| Dairy | | | Removal of H₂O₂ from milk | | | | | Removal of lactose from milk and whey | Cheese ripening, ice cream manufacture | | Cheese production, protein hydrolysates stabilization of evaporated milk |

| | | | | | | | | | |
|---|---|---|---|---|---|---|---|---|---|
| *Eggs dried* | | | | | | Glucose removal | Improved whipping and emulsification | | Improved drying |
| *Flavours* | Clarification, malt syrups | | | | Extraction | Oxygen removal | Fat extraction | Extraction | Malt syrups, savoury flavours |
| *Fruit juice* | Improved extraction | Processing | | | Extraction | Oxygen removal | | Clarification, gel prevention, yield improvement | |
| *Laundry* | Detergents | | | | | | | | Detergents |
| *Leather* | | | | | | | | | Dehairing, bating |
| *Meat* | | | | | | | Detergents | | Tenderization, scrap recovery, fish protein concentrates |
| *Photography* | | | | | | | | | Recovery of silver from spent film |
| *Rubber* | | | | Formation of foam rubber | | | | | Production |
| *Soft drinks* | | | Stabilization of citrus terpenes | | | Stabilization of citrus terpenes | | | |
| *Textiles* | Desizing, production of paper pastes and binding agents | | | | Desizing, paper modification, textile printing | | | | |
| *Vegetables* | Liquification of purees and soups | | | | Extraction | | | Hydrolysate production | |
| *Wine* | Production of must | | Colour control | Production | Clarification | Colour control | | Must production, clarification | |

**Table 14.6** Current uses of immobilized enzymes

| Enzyme | Product | Substrate | Size of operation | Immobilization method | Operational life | Reactor | Productivity |
|---|---|---|---|---|---|---|---|
| Aminoacylase (Poulsen, 1984) | L-Amino acid | DL-Acylamino acid | <250 tonnes L aa year$^{-1}$ <5 tonnes enzyme year$^{-1}$ | Absorption on DEAE sephadex | 3–4 weeks | Column | |
| Amyoglucosidase glucoamylase (Poulsen, 1984) | Glucose | Dextrins (from starch) | <5000 tonnes glucose year$^{-1}$ <1 tonne enzyme year$^{-1}$ | Absorbed to charcoal | must be >5 weeks | Column or stirred tank | Dubious economics |
| Glucose Isomerase (Xylose Isomerase) (Poulsen, 1984) | High fructose syrup (45% fructose) (45% glucose) | Glucose syrup | $3.6 \times 10^6$ tonnes 50% Fructose syrup year$^{-1}$ 1600 tonnes enzyme year$^{-1}$ | Usually cell source glutaraldehyde cross-linked | 9–17 weeks (25% activity left at end) | Column | 2000–2200 kg product/kg enzyme |
| Hydantoinase (Poulsen, 1984) | for example, D-Phenylglycine | Hydantoinglycine | <50 tonnes D-phenylglycine year$^{-1}$ 1 tonne enzyme year$^{-1}$ | Not known | — | — | — |
| Lactase (β-galactosidase) (Poulsen, 1984) | Glucose and galactose | Lactose (chiefly whey) | <1000 tonnes lactose hydrolysates year$^{-1}$ <5 tonnes enzyme year$^{-1}$ | Covalent to silica; absorption onto resin; cellulose acetate entrapment | Several weeks (<40 batch operations) | Column or tank | Dubious economics |

| Enzyme (source) | Product | Substrate | Production rate | Immobilization method | Operational stability | Reactor | Economics |
|---|---|---|---|---|---|---|---|
| Nitrilase (Poulsen, 1984) | Acrylamide | Acrylonitrile | <5 tonnes acrylamide year$^{-1}$ <0.1 tonnes enzyme year$^{-1}$ | Entrapment in cationic acrylamide gel | Not published | Column | Dubious economics |
| Penicillin G acylase (Poulsen, 1984) | 6-Amino penicillanic acid | Penicillin G (and cephalosporins) | 4000 tonnes 6-APA year$^{-1}$ 3.5 tonnes enzyme year$^{-1}$ | Covalent binding to Sephadex | 18 weeks | Column | 1500 kg product/kg enzyme |
| | | | | Cells bound to glutaraldehyde activate polyacrylonitrile | 7 weeks | Column | 600 kg product/kg enzyme |
| Penicillin V acylase (Poulsen, 1984) | 6-amino penicillanic acid | Penicillin V | 500 tonnes 6-APA year$^{-1}$ 1 tonne enzyme year$^{-1}$ | Cells cross-linked with glutaraldehyde | 9 weeks | Column or stirred batch | 350 kg product/kg enzyme |
| Keto acid dehydrogenase and formate dehydrogenase (Wandrey, 1984) | L-Methionine L-Valine L-Phenylalanine | Respective keto acids | 150 tonnes year$^{-1}$ | Soluble enzymes and polyethylene glycol–NAD | — | Membrane reactor | — |
| Fumarase (Brevibacterium flavum cells) (Chibata, Tosa and Takata, 1983) | L-Malic acid | L-Fumaric acid | 360 tonnes year$^{-1}$ 2 tonnes enzyme year$^{-1}$ | Entrapment of whole cells in κ-carrageenan and polyethyleneimine | Approx. 15 weeks | Column | 70% of maximum theoretical yield 150 kg product/kg cells |
| Aspartase E. coli (Hamilton et al., 1985) | L-Aspartate | L-Fumarate | — | Various, such as absorbed to vermiculite | Up to 6 months | Column | 100% yield 4000 kg product kg/enzyme kg |

the action of L-amino acid acylase. A process based on the use of the soluble enzyme was introduced into Japan in the early 1960s. It was a simple batch process where D,L-acylamino acid was stirred in a tank with the enzyme and at the end of the reaction the enzyme denatured and the L-amino acid separated from the D-acylamino acid. Towards the end of the 1960s the Tanabe Seiyaku Company developed a process based upon the use of the immobilized enzyme. Over 40 different immobilization procedures were employed and of these just three were selected for further evaluation, entrapment within a polyacrylamide gel, covalent bonding to iodoacetyl cellulose, and electrostatic bonding to DEAE Sephadex. Table 14.7 lists the major characteristics of these three preparations. It is clear from this data that there were no major differences in the effective characteristics of these preparations, although the higher values (compared to the soluble enzyme) of the maximum velocities of the immobilized preparations is a little perplexing (see Section 14.2.4.5). The only major difference was in their stability to heat, the DEAE Sephadex appearing the most stable. This aside, what then were the considerations that led to the final selection of the DEAE Sephadex enzyme? Its potential disadvantages were the relative lability of the enzyme–polymer bond and the high cost of the DEAE Sephadex. However, preparation of the immobilized enzyme was simplicity itself. In addition, the potential disadvantage, lability of the enzyme–polymer bond, was turned to practical advantage, because it allowed the easy regeneration of the activity of the immobilized

**Table 14.7** Summary of enzymic properties of various immobilized amino acid acylases. (Reproduced with permission from Chibata and Tosa, 1976)

| | | Immobilized amino acid acylase* | | |
| | | DEAE- | Iodoacetyl | |
| *Properties* | *Native** | *Sephadex* | *cellulose* | *Polyacrylamide* |
| --- | --- | --- | --- | --- |
| Optimum pH | 7.5–8.0 | 7.0 | 7.5–8.0 | 7.0 |
| Optimum temperature | 60 °C | 72 °C | 55 °C | 65 °C |
| Activation** energy | | | | |
| (kcal/mol) | 6.9 | 7.0 | 3.9 | 5.3 |
| Optimum $Co^{2+}$ (mmol l$^{-1}$) | 0.5 | 0.5 | 0.5 | 0.5 |
| $K_m$ (mmol l$^{-1}$)** | 5.7 | 8.7 | 6.7 | 5.0 |
| $V_{max}$ (mol/h)** | 1.52 | 3.33 | 4.65 | 2.33 |
| Heat stability† | | | | |
| 60 °C, 10 min (%) | 62.5 | 100 | 77.5 | 79.5 |
| 70 °C  10 min (%) | 12.5 | 87.5 | 62.5 | 34.5 |
| Operational‡ | | | | |
| stability (days) | — | 65 days @ 50 °C | — | 48 days @ 37 °C |
| Preparation | — | Easy | Difficult | Medium |
| Enzyme activity | — | High | High | High |
| Cost of immobilization | — | Low | High | Moderate |
| Binding force | — | Medium | Strong | Strong |
| Operational stability | — | High | — | Moderate |
| Regeneration | — | Possible | Impossible | Impossible |

\* Substrate acetyl-D,L-methionine
\*\* Assayed at 37 °C and pH 7.0
† Remaining activity
‡ Time required for 50% enzyme activity to be lost

enzyme, without the necessity of removing the DEAE Sephadex from the reactor. In operation a $1000 \, dm^3$ column was packed with DEAE Sephadex enzyme and at a flow rate of 2.8 space velocities per hour at $50 \, °C$ (that is, 2.8 times the reactor void volume/h) $2000 \, dm^3$ of $0.2 \, mol \, dm^{-3}$ acetyl D,L-methionine/h was 100% converted to acetyl-methionine and L-methionine. The activity of the reactor slowly declined to 60% of the original value after 30 days operation, whereupon its activity was regenerated by adding fresh enzyme to the reactor. The DEAE Sephadex turned out to be an extremely stable polymer matrix; even after 10 years of operation the DEAE Sephadex had not been changed! The product mixture was evaporated, the L-methionine crystallized, the acetyl-D-methionine racemized by heating to $60 \, °C$ with acetic anhydride and fed back into the substrate stream. Total yield was 91% of the theoretical maximum.

Comparison of the economics of the immobilized and soluble enzyme processes showed that the immobilized process was approximately half the cost. The savings result from three major factors: enzyme cost, understandably lower for the immobilized process; fully automatic continuous processing reducing labour costs, although greater operator skill was required, the higher yields possible resulting in lower substrate costs. Energy costs were marginally higher, because of the extra power requirement of automation, and there was the additional cost of the DEAE Sephadex. This latter had, however, disappeared from the calculations within a few years of the operation beginning. Table 14.8 summarizes these findings. Neither analysis takes account of capital costs which are higher for the immobilized enzyme process.

There have been many other propositions for the use of immobilized enzymes, most of which have been developed only on a laboratory or occasionally pilot plant scale, largely because of their dubious economics. It would be impossible to review all of them here, so we must satisfy ourselves with a brief mention of but a few, selecting those which are most likely to succeed in the future. Interestingly they all have one thing in common, the use of immobilized cells rather than purified enzymes.

A process for the conversion of sucrose to isomaltulose has been patented by Tate and Lyle (Cheetham, Imber and Isherwood, 1982). Isomaltulose has been proposed as a useful non-cariogenic bulking agent for the food and pharmaceutical industries, for which its use is being evaluated. The cynical might suggest that it is a classic case of a product without a market. The process is based on the use of a packed bed reactor filled with alginate entrapped cells of *Erwinia rhaptonica*. The particularly interesting

**Table 14.8** Economic comparison of conventional soluble batch and continuous immobilized L-amino acid acylase processes.

| Component | Relative cost | |
| --- | --- | --- |
| | *Soluble* | *Immobilized* |
| Substrate | 0.52 | 0.40 |
| Enzyme | 0.25 | 0.01 |
| Labour | 0.20 | 0.06 |
| Fuel | 0.03 | 0.05 |
| DEAE Sephadex | — | 0.02 |
| Total | 1.00 | 0.54 |

aspect of this process lies in the use of very high substrate concentrations (up to $1.6\,mol\,dm^{-3}$) which markedly stabilized the immobilized but not the free cells which had an operational half-life in excess of 1 year at $30\,°C$, and which also allowed easy product recovery.

Among the more unusual conversions which have been proposed are two uses of gel entrapped immobilized cells of *Nocardia* spp. In the first case Wingard *et al.* (1985) report the use of such cells to catalyse the epoxidation of propylene (see Section 15.3), in the second the conversion of cortexolone to cortisol (an $11\beta$-hydroxylation). This latter conversion has also been achieved using alginate entrapped spores of *Curvularia lunata* (Ohlsen *et al.*, 1980). In each case the cells were in a viable but non-growing state. The use of such biocatalysts operating on gaseous or water-insoluble compounds is where much of the future potential of the application of immobilized biocatalysts lies.

## 14.8   Summary

The successful application of an enzyme's catalytic abilities depends upon a number of factors. Crucial to this are questions of enzyme stability, reusability and ease of removal from the reaction, the lack of application of an enzyme often due to it not performing well enough in one of these areas. However, it is possible to enhance an enzyme's characteristics by a number of different methods.

The principal such method is that of enzyme immobilization, the fixing of an enzyme in some way onto a polymeric support matrix. A large number of different methods of immobilization are available, broadly classified into groups depending upon the type of bond holding enzyme and polymer together. Choice of an appropriate method is largely empirical. Immobilization may affect the enzyme's behaviour either directly, by modifying the intrinsic characteristics, or indirectly by providing a heterogeneous microenvironment around the enzyme. Thus the stability, $K_m$, maximum velocity, susceptibility to pH changes, activators or inhibitors of an enzyme may all be altered; alterations may be real intrinsic changes or merely apparent due to the imposition of a defined microenvironment. Similar effects may be observed when whole cells are immobilized. Other methods of enhancing an enzyme's characteristics also exist or are being developed. For example, selection of enzymes from thermophilic organisms, site-directed mutagenesis for the alteration of tertiary structure, modifying the reaction environment, direct chemical modification of the enzyme, are all techniques with potential. However, in all these cases there is a need for a better understanding of the structure/function relationship in enzymes, if attempts to improve them are to be other than largely empirical.

When considering the application of a biocatalyst, the question whether to use soluble or immobilized cells or enzymes is usually answered by consideration of the actual process envisaged and the relative costs of such choices.

Where enzymes are to be employed in a reactor, careful consideration must be given to the design and operation of the reactor. The performance characteristics associated with a particular design and their interaction with the physical and kinetic characteristics of the enzyme affect the yield and productivity of a process.

Enzymes find application in therapeutic, analytical, manipulative and industrial processes, full discussion of which is beyond the scope of this book. They are, however, an integral part of modern life and their range of applications will surely be extended in the future.

# Chapter 15

## *Problems and perspectives*

Enzyme technology is a rapidly developing field, and like all such branches of science, holds many promises for the future, which if their full potential is to be realized demand solutions to a number of pressing problems. Thus we shall turn our attention to some of these potential developments and, in our discussion, highlight the approaches being taken. Predicting the future is an uncertain art at the best of times, and by the time these words appear in print, events may have overtaken some of these predictions.

## 15.1 Coenzyme requiring reactions

By far the largest number of enzymes in nature are the oxidoreductases, closely followed by the kinases. It should be no surprise, however, that few of these enzymes have found large-scale industrial application and that the majority of enzymes used on a large scale are simple hydrolases. The reason is simply the requirement of these enzymes for a continual supply of expensive coenzyme. The usefulness of these enzymes is not in doubt. For example, the NAD(P) linked dehydrogenases can catalyse an extensive range of reactions: alcohol or amine to carboxyl; oxidative decarboxylations; aldehyde to carboxylic acid; amine to imine; alkane to alkene; most of these are freely reversible. In addition, these reactions are carried out with absolute stereospecificity thus making dehydrogenases potentially valuable in the chemical synthesis of, for example, steroids or pharmacologically active compounds. Stereo-specific phosphorylations catalysed by kinases could prove equally useful. The problem, apart from the inherent instability of the enzyme, is how to reduce the cost of the coenzyme. The approach is two-fold; recycle the coenzyme and retain it in the reactor.

### 15.1.1 *Coenzyme recycling*
In theory coenzymes may be recycled in a number of ways, chemically, electro-chemically or enzymically. The method of choice seems to be enzymic and all manner

of schemes have been proposed (Fig. 15.1). The principle of recycling is simple enough; add to the system a suitable second enzyme/substrate combination capable of undertaking the reverse reaction. Proposals for recycling ATP are often confounded by their economics and the high free energy release (that is, the irreversibility) of phosphoryl transfer reactions. Even where proposals have been realized, none, as yet, has found practical application and the future for recycled ATP is uncertain. More success has been achieved with the recycling of NAD/NADH. The system of choice for NADH regeneration, the usual requirement, is to employ formate dehydrogenase and formic acid. The distinct advantage of this system is that the products are the desired NADH and carbon dioxide, which may be easily removed from the reactor without employing complex separations. Until recent times, however, the cost of producing formate dehydrogenase was prohibitive, but use of two-liquid phase separation has dramatically reduced its price.

Detailed discussion of chemical or electrochemical regeneration of coenzymes is beyond the scope of this text. It will suffice to point out that chemical redox compounds, for example phenazine methosulphate or 2,6-dichloroindophenol, may be difficult to separate from the reaction mixture. Equally electrochemical coenzyme recycling requires efficient electron transfer between any mediator, the coenzyme or even the enzyme and the electrode surface and on a large scale this may be difficult to achieve (see Jones and Taylor, 1976; Lowe, 1981; and Wingard, Shaw and Castuer, 1982).

### 15.1.2   *Coenzyme retention*

Coenzyme retention within a reactor is more problematic. Obviously the small size of coenzyme molecules means that they will readily escape from an enzyme reactor unless they are first modified in some way.

This modification may take one of three forms. The coenzyme may be immobilized by covalently bonding it onto an insoluble polymer. Clearly it can then easily be retained in a reactor. Equally obviously, however, is the fact that it is unlikely to interact with an immobilized enzyme and thus a soluble enzyme must be employed. This in turn creates the problem of how to retain the enzyme in the reactor! Thus, insoluble immobilized coenzymes are largely confined in use to affinity chromato-

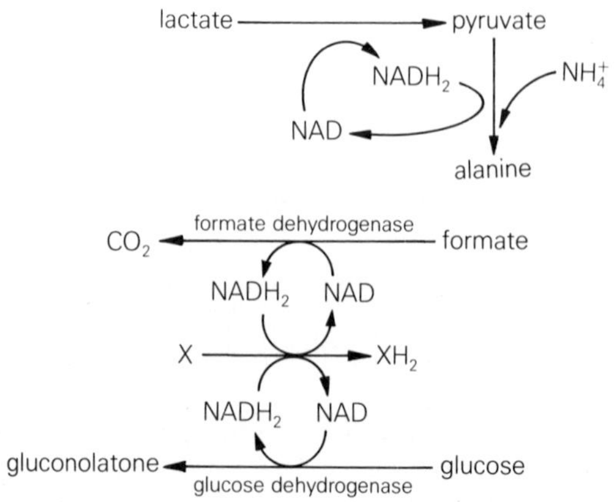

**Fig. 15.1**   Schemes for recycling NAD/NADH

graphy. There is, however, one exception, an intriguing proposal which overcomes the problem of immobilizing both the enzyme and coenzyme. The idea was to modify NAD by linking it to a suitable acrylamide derivative and incorporating it into a polyacrylamide gel which was also used to entrap the dehydrogenase enzymes. Thus the enzymes were insoluble but trapped within the NAD-bearing gel structure and hence NAD and enzyme could interact. However, even here there remains the problem of the low reactivity of immobilized coenzymes, typically less than 1% of the soluble coenzyme analogue. These considerations gave rise to the second of our three approaches, the synthesis of soluble immobilized coenzymes. NAD may be easily attached to soluble polymers, for example, dextran or polyethylene glycol, in such a way that it retains its reactivity. Enzyme and high molecular weight-modified NAD can then be coimmobilized by gel entrapment, or simply retained in the reactor by ultrafiltration. The use of hollow fibre reactors is particularly pertinent here. The third approach is perhaps the simplest, to immobilize the coenzyme directly onto the dehydrogenase. As unlikely as it sounds, by judicious use of a spacer molecule between the coenzyme and enzyme, this approach can be made to work. Mansson, Larsson and Mosbach (1978, 1979) describe the preparation of an NAD $N^6$,(6-amino-hexyl)carbamylmethyl-lactate dehydrogenase complex where the NAD was available (and active) to the lactate dehydrogenase and also available to a recycling enzyme.

How, then, can NAD, or for that matter ATP or coenzyme A (CoA), be linked successfully to a polymer matrix? A variety of methods have been developed and, in general, these show that direct linking of the coenzyme to the polymer results in preparations with negligible activity. Thus a spacer molecule has to be included between the coenzyme and polymer. This is best achieved by building a suitably reactive linear molecule onto the coenzyme and then bonding this derivatized coenzyme onto the polymer via a reactive group (usually an amine) on the free end of the spacer arm. The two most commonly used spacer molecules are carboxymethyl and (6-aminohexyl)carbamylmethyl groups. Care must be taken over where on the coenzyme the spacer molecule is attached, in order to retain maximum activity. For NAD(P), attachment via the nicotinamide or ribose moiety usually results in an inactive coenzyme and the best positions are on the adenine moiety. For NAD the point of attachment is critical, only substitution on the N-6 atom of the adenine is acceptable, whether carboxymethyl or (6-aminohexyl)carbamylmethyl groups are attached. In contrast N-1 or N-6 substituted carboxymethyl-NADP derivatives are both active, whereas (6-aminohexyl)carbamylmethyl substitution at either position inactivates NADP. Clearly consideration of charge as well as substitution point is important. In addition to such observations the choice of immobilization method may also depend upon the enzyme ultimately involved. For example, NAD $N^6$,(6-amino-hexyl)carboxymethyl-dextran is an active substrate for mammalian liver alcohol dehydrogenase, but not for yeast alcohol dehydrogenase.

There are two general drawbacks to the application of these developments. First is cost: NAD analogues are expensive and tedious to produce; second is instability: the single bond between the NAD and polymer or enzyme, particularly when cyanogen bromide activation of the polymer is used, is relatively unstable and such preparations have limited operational lives.

Thus there are many problems still to be overcome in this field before coenzyme requiring enzymes can find practical application as productive catalysts. Two other approaches however are under investigation/use as a way of circumventing these problems. The first is to use whole cells for dehydrogenase or kinase activity (see Section 14.4); the second is the search for flavoprotein or quinoprotein dehydrogenases (for example, methanol dehydrogenase from certain sources) which operate via bound prosthetic groups.

## 15.2   Oxidases and oxygenases

Not all oxidoreductase enzymes are coenzyme requiring, nor are all the interesting reactions catalysed by dehydrogenases. Two further groups of enzyme, the oxidases and oxygenases, also hold much promise.

### 15.2.1   *Oxygenases*

Oxygenases can introduce either one or both atoms of an oxygen molecule onto a carbon center (mono-oxygenases and dioxygenases); to date it is the mono-oxygenases that have received most attention. Among the reactions they can catalyse are: the conversion of alkanes to alcohols; olefins to epoxides; sulphides to sulphoxides; oxidative demethylation; cleavage of aromatic rings; and, perhaps most significantly, the hydroxylation of aromatic or polycyclic compounds and steroids. For example, a variety of micro-organisms, notably *Rhizopus arrhizus* and *Nocardia* spp., contain a mono-oxygenase capable of hydroxylating progesterone in the 11-$\alpha$ position. This enzyme, microbial $\Delta^1$ dehydrogenases and some chemical steps have been used for a number of years, for the conversion of progesterone to hydrocortisone, prednisolone, cortisone and prednisone. Until recently the process has been one of fermentation, but attempts are now being made to use immobilized cells (principally *Nocardia* spp.) to perform the 11-$\alpha$ hydroxylations. For further discussion of this, the reader is referred to a series of papers by various authors in Chibata, Fukin and Wingard (1982).

Clearly there is both much potential and money in the application of enzymes to steroid synthesis, ultimately perhaps the total enzymic conversion of cholesterol to a whole range of steroids.

### 15.2.2   *The oxidases*

The oxidases are an equally useful group of enzymes, which include flavoproteins (such as amino acids and glucose oxidases) metaloflavoproteins (such as xanthine oxidase) and haem-based proteins (such as catalase and peroxidases); some of these enzymes have already found application. Glucose oxidase has been used for many years in the food industry, with catalase, as an antioxidant and catalase alone is used in the production of foam rubber. More recent applications proposed included the production of fructose from glucose using pyranose-2-oxidase (EC 1.1.3.10) to convert glucose to glucosone followed by chemical reduction (Geigert, Neidleman and Hirano, 1983); the production of $\alpha$-keto acids from amino acids using amino acid oxidases (Szwaher, Brodelius and Mosbach, 1982); the production of alkene oxides from alkenes using haloperoxidases (Neidleman, Amon and Geigert, 1981). This last application is based upon the incorporation of halide molecules (such as chlorine) into alkenes (such as ethylene) to give a halohydrin which is converted into the epoxide by alkali treatment and subsequently to the oxide (such as ethylene oxide). It is estimated that this enzyme, or possibly microbial, route to the alkeneoxides (important feed-stocks for polymer synthesis) could be in economic operation by 1990.

## 15.3 Non-aqueous systems

Many of the possible reactions described above involve substrates or products which are either sparingly soluble or insoluble in water. In order to facilitate economic production such conversions would best be carried out in organic solvents and thus the whole problem of enzyme stability and action in such solvents must be resolved (see Section 14.5). There are, however, a number of other positive effects of introducing organic solvents into enzyme reaction systems. When water is replaced by an organic solvent the effect is a decrease in water concentration. As we have seen, this may have two results, stabilization of the enzyme (see Section 14.5) and alteration of the equilibrium position of hydrolytic reactions favouring net synthesis. For example, α-chymotrypsin in chloroform will synthesize N-acetyl-L-tryptophan ethyl ester from ethanol and N-acetyl-L-tryptophan. Selection of a suitable solvent is important; the yield of N-benzoyl-L-phenylalanine ethyl ester by α-chymotrypsin varies from 80% in chloroform, 63% in carbon tetrachloride to 26% in diethyl ether (Martinek and Semenov, 1981). Perhaps of more immediate application is the synthesis of specific glycerides from glycerol and the appropriate fatty acid by reversal of the normal hydrolytic reaction catalysed by lipase. Thus fats resembling expensive cocoa butter can be produced from cheap substrates like palm oil.

Finally there exists the possibility that the catalytic nature of the enzyme itself may be altered by the presence of an organic solvent, presumably by modifying the structure of the active site. Pal and Gertler (1983) showed that the enzyme α-thrombin, which normally exhibits esterase and amidase activity, was affected by the presence of dimethyl sulphoxide. At a v/v concentration of 20% dimethyl sulphoxide in water the amidase activity is reduced 10-fold whilst the esterase activity is doubled.

Thus the effects of organic solvents can be unexpected and potentially very useful and will undoubtedly find increasing successful application, given proper optimization of the conditions of use. For example, organic solvents come in two basic varieties, water-miscible or immiscible. While the immiscible solvents may potentially at least be less damaging to the enzyme contained in the aqueous phase, the water-miscible solvent may deliver more of the substrate to the enzyme. This is illustrated by the work of Antonini, Arrea and Cremonesi (1981) on the yield of reduced cortisone formed from cortisone, by the action of 20β-hydroxysteroid dehydrogenase (EC.1.1.1.53). In water the conversion of the cortisone is less than 5%, in chlorobenzene 15%, while in butyl acetate it is 100%, despite the lack of inhibitory effect by chlorobenzene and the 52% inhibition of the enzyme recorded for butyl acetate. The difference may be explained on the basis of the relative solubility of cortisone in the two solvents and the solvents in water; cortisone is more soluble in butyl acetate which is more soluble in water. Thus the butyl acetate both delivers more cortisone into the aqueous phase while having a greater capability to denature the enzyme.

## 15.4 Energy production

Biological organisms depend for their very existence on energy production. It is not surprising, therefore that man has attempted over the centuries to divert this energy production for his own use. Fossil fuels, or wood, are the prime cases, utilization of biologically fixed and concentrated carbon as a source of energy. However, as future sources of energy for man's need their availability and cost are likely to make them unattractive. In more recent times, then, many different attempts have been made deliberately to subvert biological metabolic processes to provide a usable form of energy. Two basic forms of energy are available: transportable fuels and direct generation of electrical potential.

### 15.4.1  *Transportable fuels*
Three forms of transportable fuels have been proposed from biological sources: hydrogen, organic gases, and organic liquids or solids.

The principal organic liquid proposed as a fuel is, of course, ethanol. Ethanol production by fermentation is already a refined art, but it is rarely feasible to produce directly from fermentation a solution of more than 20% ethanol. Unfortunately, as anyone dropping a match into a glass of fortified wine knows, such mixtures are strictly non-combustible. Thus to produce a combustible fuel the ethanol must be distilled. Here, the laws of thermodynamics take over, for the energy input required to distil the ethanol is actually greater than the energy content of the ethanol! That the Brazilian ethanol-petroleum fuel mentioned in Chapter 1 is seen as being economic is due to three factors. First, the waste bagasse from the sugar cane from which the ethanol is produced is used to fire the stills (that is, the energy for distillation is very cheap, if not actually free). Second, in the Brazilian case, the national economy is almost totally dependent on imported oil at great cost to its foreign exchange; any means of alleviating this is viewed as politically expedient. Third, sugar as a product in its own right has a relatively low value at present. Thus in Brazil in the first half of the 1980s, 'gasahol' makes political and economic (just) sense. Were Brazil to find large reserves of oil or natural gas, the world sugar price steeply to increase, or crude oil prices to fall, the situation could be very different. However, in general terms and for the forseeable future, ethanol as a fuel appears an uneconomic proposition, unless high-yielding micro-organisms capable of producing more concentrated ethanol can be found.

Of the organic gases the most easily obtained fuel is methane. It is usually produced as a result of anaerobic metabolism from a wide range of micro-organisms (see Appendix to Chapter 3). At its simplest, the covered hole in the ground filled with animal wastes; methane production represents an ideal, simple form of biotechnological fuel production, and it is in this application that most methane fuel is likely to be produced. However, there is no theoretical reason why controlled fermentation, from waste organic material (such as lactose), could not in future produce large quantities of methane. Before this high technology is realized much work must be done on understanding, improving and controlling the methane-producing organisms. The future is perhaps brighter here than for ethanol, because of the obvious ease of separation and collection of the methane. This last point illustrates the importance of product purification and recovery costs to the feasibility of biotechnological processes.

Hydrogen is perhaps the ideal fuel. It is light, the technology for using it exists and in many ways is safer than petroleum and, if derived from the lysis of water, is pollution-free and regenerates its source on combustion. It has been known for some years that various micro-organisms (such as *Clostridium butyricum*, and various cyanobacteria) can produce molecular hydrogen when 'fed' with a reduced electron carrier (for example, ascorbate, ferredoxin or methyl viologen). The effect is a result of the action of the enzyme hydrogenase which catalyses the reduction of hydrogen ions to molecular hydrogen. The reduced electron carrier can come from a variety of sources, but a popular approach is to use the photosynthetic process of plants to provide the reducing power. Thus the plant chloroplast is illuminated, water is split to oxygen, hydrogen ions and the electrons carried through the two photosystems to reduce, for example, ferredoxin. The ferredoxin is reoxidized by the hydrogenase with the concomitant release of hydrogen (Fig. 15.2). The attraction is obviously that the energy source is light that is freely available. However, before biophotolytic hydrogen production displaces the oil industry several problems remain. The first is the instability of the hydrogenase enzyme and its inhibition by oxygen, one of the products of the reaction. This instability has in part been overcome by selecting less oxygen-sensitive enzymes and by immobilization (see Section 14.5). The second problem, instability of plant photosystems has proved more intractable. Isolated chloroplasts remain very unstable and most research has centred on the use of whole cells, particularly unicellular algae. At present such systems remain stable enough to produce hydrogen for up to 10 days. Third is the low efficiency of conversion of solar energy achieved by biological photosystems, typically 1%; unless this figure can be raised to around 10–15% biophotolytic reactors will have to be excessively large to produce significant quantities of hydrogen. Clearly then, practical application of such systems is still a long way off.

### 15.4.2 *Fuel cells*

The use of hydrogen as a fuel has been criticized on the grounds of its low energy value even when compressed. It can, however, be used to produce an electrical current by oxidation at the anode of a conventional fuel cell. This is the essence of the biofuel cell. However systems based upon the electrochemical oxidation of microbially produced hydrogen have yet to produce a net yield of energy, because of the power requirement for heating, stirring and gassing the microbial cells.

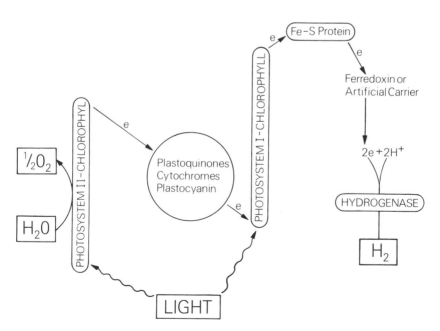

**Fig. 15.2** Reaction sequences for $H_2$ liberation by photosystems and hydrogenase

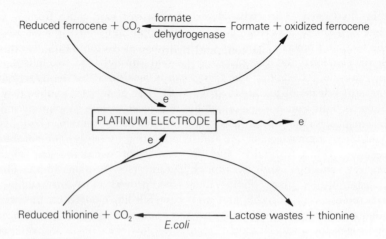

**Fig. 15.3**  Two possible configurations of biofuel cells

Research into biofuel cells probably dates from the work of Cohen (1931) who described a bacterial cell which, when coupled to an electrode by ferricyanide, acted as an electrical half-cell. In principle, all biofuel cells operate in the same way. A biocatalyst is used to reduce a material which is then oxidized at an anode. The biocatalyst may be a whole cell or an enzyme (for example, formate dehydrogenase), the reduced material may be a metabolic product or an artificial electron carrier, the oxidized material of the redox pair may be a feedstock for the biological cell or a substrate of the enzyme. Figure 15.3 illustrates some of the possibilities. Typically potentials of 40 V can be generated with up to 90% thermodynamic efficiency, but current densities are limited; the largest current by such a device reported in recent literature is $200 \, mA/m^2$ of platinum electrode (Davis *et al.*, 1983). Nevertheless for specialist applications biofuel cells may yet prove practical; low power consuming microcomputers, noiseless battery rechargers for military use, power provision for cardiac pacemakers have all been proposed. However, before even these systems can be developed, the general problem of biocatalyst instability must be overcome and methods found for either keeping the anode compartment anaerobic or preventing transfer of electrons from the reduced material to oxygen rather than the anode. For a fuller review the reader is referred to Aston and Turner (1984).

## 15.5  New reactions for old

While it has been suggested that only some 10% of the enzymes in nature have yet been characterized, it has also been suggested that of the enzymes presently known only some of their reactions have been described. Thus is borne the concept of new reactions from old enzymes. This is still a highly speculative novel field of biocatalyst technology and we can study only a few existing examples. However the potential for such a field is enormous, for it is far simpler to use an existing enzyme than to pursue a new one.

Perhaps the classic example of an enzyme being put to good use catalysing the 'wrong' reaction is in the industrial conversion of glucose to fructose by the enzyme commonly called glucose isomerase. In fact the enzyme is strictly a xylose isomerase which can function with glucose.

More recently an enzymic system for producing hydroquinone from benzoquinone has been proposed. The enzyme is glucose oxidase which would normally convert glucose and oxygen to gluconolactone and hydrogen peroxide. However, the enzyme works quite happily with benzoquinone as the oxidant, producing hydroquinone.

Substrate concentration can affect the product's structure in certain enzyme-catalysed reactions. For example, chloroperoxidase-catalysed halogenation of allyl alcohol produces monobromopropanediol with a concentration of $17\,mmol\,dm^{-3}$ potassium bromide, but dibromopropanol with $3.4\,mol\,dm^{-3}$ potassium bromide (Neidleman and Geigert, 1983).

Carboxypeptidase A has two activities, esterase and peptidase. The enzyme naturally contains an atom of zinc. Replacing this zinc with cobalt doubles the peptidase activity and increases the esterase by 14%, whereas cadmium increases the enzyme's esterase activity by 43% while reducing the peptidase activity to zero (Vallee, 1980). The substrate specificity of aminoacylase from *Aspergillus oryzae* is dependent on the metal ion present, cobalt enhances N-chloroacetyl-leucine deacylation, whilst zinc enhances the deacylation of N-chloroacetyl norleucine.

From these few examples it should be clear that much remains to be discovered about enzyme catalysis, its control and its application.

## 15.6  Summary

While it is difficult to predict the future, a number of problems must be overcome if the use of enzymes is to be extended significantly.

At present, apart from analytical use, coenzyme requiring enzymes do not have many applications, despite the undoubted value of the reactions they catalyse. Although some methods have been developed to recycle and retain coenzymes within reactor systems, the economics of most are dubious. The use of cells containing enzymes, coenzymes and recycling systems readymade appears to have a more optimistic future in this area.

Two groups of enzymes which are at present the subject of intensive study are the oxygenases and oxidases. Many such enzymes have been found to catalyse some fairly novel reactions, for example the epoxidation of alkenes, and if this technology was successfully developed it could have a major impact on what is presently thought of as the chemical industry.

The use of enzymes in non-aqueous systems was for many years regarded as an impossible proposition. However, it has been well demonstrated that enzymes will work in the presence of organic solvents, often with useful result. For example, systems containing little water may enhance an enzyme's stability or be used to reverse a normally hydrolytic reaction. Clearly, knowledge of such opeations is vital if enzyme modification of substrates sparingly soluble in water is to be carried out on a large scale.

Enzyme technology has also been turned towards energy production, with the construction of systems producing hydrogen or acting as fuel cells. It may, however, be some time before such processes become economically feasible.

One of the most exciting challenges of enzyme technology today lies in the characterization of just what reactions an enzyme can catalyse. The extension of the use of an existing enzyme to another type of reaction may indeed be a more viable option than the search for a new enzyme.

# Glossary

## E. C. Classification

| Enzyme | Recommended name (if different) | E C Number |
|---|---|---|
| Acetyl cholinesterase | | 3.1.1.7 |
| Aconitase | (Aconitate hydratase) | 4.2.1.3 |
| Adenosine deaminase | | 3.5.4.4 |
| Alcohol dehydrogenase | | 1.1.1.1 |
| Alcohol oxidase | | 1.1.3.13 |
| Aldolase | (Fructose-bisphosphate aldolase) | 4.1.2.13 |
| Alkaline phosphatase | | 3.1.3.1 |
| Amine oxidase | | 1.4.3.6 |
| L-Amino-acid oxidase | | 1.4.3.2 |
| Amino-acylase | | 3.5.1.14 |
| α-Amylase | | 3.2.1.1 |
| β-Amylase | | 3.2.1.2 |
| Amyloglucosidase | (Exo-1,4,α-D-glucosidase) | 3.2.1.3 |
| Asparaginase | | 3.5.1.1 |
| Aspartase | (Aspartate ammonia-lyase) | 4.3.1.1 |
| Aspartate aminotransferase | | 2.6.1.1 |
| Aspartokinase | (Aspartate kinase) | 2.7.2.4 |
| Bromelain | | 3.4.22.4 |
| Carboxypeptidase A | | 3.4.17.1 |
| Catalase | | 1.11.1.6 |
| Cellulases | | 3.2.1.4 |
| Chloroperoxidase | (Chloride peroxidase) | 1.11.1.10 |
| Cholesterol esterase | | 3.1.1.13 |
| Cholesterol oxidase | | 1.1.3.6 |
| α-Chymotrypsin | (Chymotrypsin C) | 3.4.21.2 |
| Citrate condensing enzyme | (Citrate (si)-synthase) | 4.1.3.7 |
| Cysteinyl-tRNA synthetase | | 6.1.1.16 |

| Enzyme | Recommended name (if different) | E C Number |
|---|---|---|
| Dimethylamine dehydrogenase | | 1.5.99.7 |
| DNA ligase | (Polydeoxyribonucleotide synthetase (ATP)) | 6.5.1.1 |
| DNA polymerase | (DNA nucleotidyl transferase) | 2.7.7.7 |
| Endonucleases–see Restriction Endonucleases | | 3.1.23 |
| Enolase | | 4.2.1.11 |
| Exonucleases | (Exodeoxyribonucleases) | 3.1.11 |
| Ficin | | 3.4.22.3 |
| Formaldehyde dehydrogenase | | 1.2.1.1 |
| Formate dehydrogenase | | 1.2.1.2 |
| Fructose phosphate kinase | (6-Phosphofructokinase) | 2.7.1.11 |
| Fructose bisphosphate aldolase | | 4.1.2.13 |
| Fructose-1,6-bisphosphatase | (Fructose bisphosphatase) | 3.1.3.11 |
| Fumarase | (Fumarate hydratase) | 4.2.1.2 |
| Galactokinase | | 2.7.1.6 |
| $\beta$-Galactosidase | ($\beta$-D-Galactosidase) | 3.2.1.23 |
| $\beta$-Glucanase | (Exo-1,4-$\beta$-D-Glucosidase) | 3.2.1.74 |
| 1,4-$\alpha$-Glucanphosphorylase | (Phosphorylase) | 2.4.1.1 |
| Glucoamylase | (Exo-1,4,$\alpha$-D-Glucosidase) | 3.2.1.3 |
| Glucose dehydrogenase | | 1.1.1.47 |
| Glucose isomerase–name now deleted, reaction actually due to xylose isomerase (5.3.1.5) or glucose phosphate isomerase (5.3.1.9) in the presence of arsenate. | | (5.3.1.18) |
| Glucose oxidase | | 1.1.3.4 |
| Glucose-6-phosphate dehydrogenase | | 1.1.1.49 |
| Glutamate dehydrogenase | (Glutamate dehydrogenase (NAD(P)$^+$)) | 1.4.1.3 |
| Glutamate synthetase | (Glutamate synthease (NADPH)) | 1.4.1.13 |
| Glutamine synthetase | | 6.3.1.2 |
| Glyceraldehyde-3-phosphate dehydrogenase | | 1.2.1.12 |
| Glycerokinase | (Glycerate Kinase) | 2.7.1.31 |
| Glycollate dehydrogenase | | 1.1.99.14 |
| Glycollate oxidase | | 1.1.3.1 |
| Haloperoxidase–see chloroperoxidase | | |
| Hexose phosphate isomerase | (Glucose phosphate isomerase) | 5.3.1.9 |
| Histidine ammonia-lyase | | 4.3.1.3 |
| Homoisocitrate lyase | (Homocitrate synthease) | 4.1.3.21 |
| Hydantoinase | (Dihydropyrimidinase) | 3.5.2.2 |
| Hydrogenase | | 1.18.3.1 |
| 21-$\Delta$-Hydroxylase | (21-$\Delta$-Hydroxysteroid dehydrogenase) | 1.1.1.150 |
| Hydroxypyruvate reductase | | 1.1.1.81 |
| 20$\beta$-Hydroxysteroid dehydrogenase | | 1.1.1.53 |

| Enzyme | Recommended name (*if different*) | E C Number |
|---|---|---|
| Invertase | ($\beta$-D-Fructofuranosidase) | 3.2.1.26 |
| Isocitrate dehydrogenase | (Isocitrate dehydrogenase (NADP$^+$)) | 1.1.1.42 |
| Isocitrate lyase | | 4.1.3.1 |
| Isoleucine-tRNA synthetase | | 6.1.1.5 |
| 2-Keto-3-deoxy-6-phospho-gluconate aldolase | (Phospho-2-keto-3-deoxy-gluconate aldolase) | 4.1.2.14 |
| $\alpha$-Ketoglutarate dehydrogenase | (Oxoglutarate dehydrogenase lipoamide) | 1.2.4.2 |
| $\beta$-Lactamase | (Penicillinase) | 3.5.2.6 |
| Lactase | ($\beta$-D-Galactosidase) | 3.2.1.23 |
| Lactate dehydrogenase | | 1.1.1.27 |
| Leucyl-tRNA synthetase | | 6.1.1.4 |
| Lipase | (Triacylglycerol lipase) | 3.1.1.3 |
| Lipoxygenase | | 1.13.11.12 |
| Lysozyme | | 3.2.1.17 |
| Malate dehydrogenase | | 1.1.1.37 |
| Malate synthetase | (Malate synthase) | 4.1.3.2 |
| Malyl CoA lyase | | 4.1.3.24 |
| Malyl CoA synthetase | | 6.2.1.9 |
| $\alpha$-D-Mannosidase | | 3.2.1.24 |
| Methane mono-oxygenase | (Alkane 1-monooxygenase) | 1.14.15.3 |
| Methanol dehydrogenase | (Alcohol dehydrogenase (acceptor)) | 1.1.99.8 |
| Methanol oxidase | (Alcohol oxidase) | 1.1.3.13 |
| Methionine-tRNA synthetase | | 6.1.1.10 |
| Methylamine dehydrogenase | (Amine dehydrogenase) | 1.4.99.3 |
| Methylglutamate synthetase | (Methylamine-glutamate methyltransferase) | 2.1.1.21 |
| Nitrilase | | 3.5.5.1 |
| Nucleoside(-5'-)phosphtransferase | | 2.7.1.77 |
| Ornithine carbamoyltransferase | | 2.1.3.3 |
| Papain | | 3.4.22.2 |
| Pectinase | (Polygalacturonase) | 3.2.1.15 |
| Penicillin acylase | | 3.5.1.11 |
| Peroxidase | | 1.11.1.7 |
| Phospho(enol)pyruvate carboxylase | | 4.1.1.31 |
| Phospho(enol)pyruvate hydratase | (Enolase) | 4.2.1.11 |
| 6-Phosphofructokinase | | 2.7.1.11 |
| (6-)Phosphogluconate dehydratase | | 4.2.1.12 |
| (6-)Phosphogluconate dehydrogenase | | 1.1.1.44 |
| (6-)Phosphogluconolactonase | | 3.1.1.31 |
| (3-)Phosphoglycerate kinase | | 2.7.2.3 |
| Proteases | | 3.4 |
| Pullulanase | | 3.2.1.41 |
| Pyranose(-2-)oxidase | | 1.1.3.10 |
| Pyruvate carboxylase | | 6.4.1.1 |
| Pyruvate decarboxylase | | 4.1.1.1 |

| Enzyme | Recommended name (if different) | E C Number |
|---|---|---|
| Pyruvate dehydrogenase | (Pyruvate dehydrogenase (lipoamide)) | 1.2.4.1 |
| Pyruvate kinase | | 2.7.1.40 |
| Rennin | (Chymosin) | 3.4.23.4 |
| Restriction endonucleases | (Endodeoxyribonucleases) | 3.1.23 |
| Restriction endonuclease *Ava*I | | 3.1.23.3 |
| Restriction endonuclease *Bal*I | | 3.1.23.5 |
| Restriction endonuclease *Bam*HI | | 3.1.23.6 |
| Restriction endonuclease *Cla*I | | |
| Restriction endonuclease *Eco*RI | | 3.1.23.13 |
| Restriction endonuclease *Hind* III | | 3.1.23.21 |
| Restriction endonuclease *HPa*I | | 3.1.23.23 |
| Restriction endonuclease *Pst*I | | 3.1.23.31 |
| Restriction endonuclease *Pvu*I | | 3.1.23.32 |
| Restriction endonuclease *Pvu*II | | 3.1.23.33 |
| Restriction endonuclease *Sal*I | | 3.1.23.37 |
| Restriction endonuclease *Sau*3A | | 3.1.23.27 |
| Rhodanase | (Thiosulphate sulphur-transferase) | 2.8.1.1 |
| Ribulose(1-5-)bisphosphate carboxylase | | 4.1.1.39 |
| Ribulose-5-phosphate kinase | (Phosphoribulokinase) | 2.7.1.19 |
| RNA polymerase | (RNA nucleotidyltransferase) | 2.7.7.6 |
| $S_1$ nuclease | (Endonuclease $S_1$ (Aspergillus)) | 3.1.30.1 |
| Sedoheptulose bisphosphatase | | 3.1.3.37 |
| Serine-glyoxylate aminotransferase | | 2.6.1.45 |
| Serine hydroxymethyltransferase | | 2.1.2.1 |
| Subtilisin | | 3.4.21.14 |
| Succinic dehydrogenase | (Succinate dehydrogenase) | 1.3.99.14 |
| Superoxide dismutase | | 1.15.1.1 |
| Tatronic acid semialdehyde reductase | (Tartronate semialdehyde reductase) | 1.1.1.60 |
| Terminal transferase | (DNA nucleotidylexotransferase) | 2.7.7.31 |
| Thermolysin | (*Bacillus thermoproteolyticus* neutral proteinase) | 3.4.24.4 |
| $\alpha$-Thrombin | | 3.4.21.5 |
| Transaldolase | | 2.2.1.2 |
| Triokinase | | 2.7.1.28 |
| Triosephosphate isomerase | | 5.3.1.1 |
| Trypsin | | 3.4.21.4 |
| Tryptophanyl-tRNA synthetase | | 6.1.1.2 |
| Tyrosyl-tRNA synthetase | | 6.1.1.1 |
| Urease | | 3.5.1.5 |
| Uricase | (Urate oxidase) | 1.7.3.3 |
| Urocanase | (Urocanate hydratase) | 4.2.1.49 |
| Urokinase | | 3.4.21.31 |

| Enzyme | Recommended name *(if different)* | E C Number |
|---|---|---|
| Valyl-tRNA synthetase | | 6.1.1.9 |
| Xanthine oxidase | | 1.2.3.2 |
| Xylose isomerase | | 5.3.1.5 |

# References

Ajinomoto, A. (1983). European patent application 71023.

Al Obaidi, Z. S. and Berry, D. R. (1980), *Biotechnology Letters*, **2**, 5–10.

Antonini, E., Arrea, G. and Cremonesi, P. (1981). *Enzyme and Microbial Technology*, **3**, 291–296.

Aston, W. J. and Turner, A. P. F. (1984). In *Biotechnology and Genetic Engineering Reviews*, Vol. 1. Ed. Russell, G. E. pp. 89–120. Newcastle, Intercept.

Atkinson, A., Banks, G. T., Bruton, C. J., Comer, M. J., Jakes, R. J., Kamalagharan, T., Whitaker, A. R. and Winter, G. P. (1979). *Journal of Applied Biochemistry*, **1**, 247–58.

Ball, C. and MacGonagle, M. P. (1978). *Journal of Applied Bacteriology*, **45**, 67–74.

Baross, J. A. and Demming, J. W. (1983). *Nature*, **303**, 423–6.

Bascombe, S., Banks, G. T., Skarstedt, M. T., Fleming, A., Bettelheim, A. and Connors, T. A. (1975). *Journal of General Microbiology*, **91**, 1–16.

Beardsmore, A. J., Aperghis, P. N. G. and Quayle, J. R. (1982). *Journal of General Microbiology*, **128**, 1423–39.

Brewersdorff, M. and Dostalek, M. (1971). *Biotechnology and Bioengineering*, **13**, 49–62.

Bruton, C. J. (1983). *Philosophical Transactions of the Royal Society (London) B*, **300**, 249–261.

Buckland, B. C., Richmond, W., Dunnill, P. and Lilly, M. D. (1974). In *Industrial Aspects of Biochemistry*. (FEBS Vol. 30). Ed. Spencer, B., pp. 65–79. Amsterdam, North-Holland Pub. Co.

Bu'Lock, J. D., Hamilton, D., Hulme, M. A., Powell, A. J., Shepherd, D. Smalley, H. M. and Smith, G. N. (1965). *Canadian Journal of Microbiology*, **11**, 765–78.

Bu'Lock, J. D. (1980). *Biotechnology Letters*, **3**, 285–90.

Butterworth, D. (1984). In *Biotechnology of Industrial Antibiotics*, pp. 225–36. Ed. Vandamme, E. J. New York. Marcel Dekker.

Calam, C. T. and Smith, G. M. (1980). *FEMS Microbiology Letters*, **10**, 231–4.

Calcott, P. H. (1981). In *Continuous Culture of Cells*, Vol. 1, pp. 13–26. Ed. Calcott, P. H. Boca Raton, Louisiana, CRC Press.

Campbell, I. (1984). *Advances in Microbiology and Physiology*, **25**, 2–60.

Carr, P. W. and Bowers, L. D. (1980). *Immobilized Enzymes in Analytical and Clinical Chemistry*. New York, John Wiley.

Carrea, G. (1984). *Trends in Biotechnology*, **2**, 102–6.

Carrea, G., Bavara, R. and Pasta, P. (1982). *Biotechnology and Bioengineering*, **24**, 1–7.

Cashion, P., Javed, A., Harrison, D., Seeley, J., Lentini, V. and Sathe, G. (1982). *Biotechnology and Bioengineering*, **34**, 403–23.

Chang, L. T., Terasaka, D. T. and Elander, R. P. (1982). *Developments in Industrial Microbiology*, **23**, 21–9.

Chater, K. F. (1979). In *Genetics of Industrial Microorganisms* Ed. Sebek, O. K. and Laskin, A. I., pp. 123–33. Washington, American Society of Microbiology.

Cheetham, P. S. J., Imber, C. E. and Isherwood, J. (1982). *Nature*, **299**, 628–31.

Chibata, I., and Tosa, T. (1976). In *Applied Biochemistry and Bioengineering*, Vol. I. Ed. Wingard Jr. L. B., Katchalski-Katzir, E. and Goldstein, L., pp. 330–8. New York, Academic Press.

Chibata, I., Fukin, S. and Wingard, L. B. (1982) (eds) *Enzyme Engineering*, Vol. 6, pp. 117–34. New York, Plenum Press.

Chibata, I., Tosa, T. and Takata, I. (1983). *Trends in Biotechnology*, **1**, 9–11.

Clark, L. C. and Lyons, C. (1962). *Annals of the New York Academy of Sciences*, **102**, 29–45.

Cohen, B. (1931). *Journal of Bacteriology*, **21**, 18–19.

Colby, J., Williams, E. and Turner, A. P. F. (1985). *Trends in Biotechnology*, **3**, 12–17.

Colson, C., Cornelius, P., Digneffe, C., Walon, R. and Walon, C. (1981). European patent office, patent application number, 81,3005782.

Crameri, R., Davies, J. and Thompson, C. (1985). *The World Biotech Report 1985 Vol. 1: Europe. Proc. of Biotech '85, Europe, Geneva*, pp. 47–54. London, Online Publications.

Dagley, S. (1978). *The Bacteria*, Vol. 6, pp. 305–88. New York, Academic Press.

Dalton, H. (1980a). In *Hydrocarbons in Biotechnology*. Eds. Harrison, D. E. F., Higgins, I. J. and Walkinson, R. pp. 85–92. London, Heyden and Son.

Dalton, H. (1980b). *Advances in Applied Microbiology*, **26**, 71–87.

Daniels, L. (1984). *Trends in Biotechnology*, **2**, 91–98.

Davis, B. D. (1949). *Proceedings of the National Academy of Sciences, USA*, **35**, 1–10.

Davis, G., Hill, H. A. O., Aston, W. J., Turner, A. P. F. and Higgins, I. J. (1983). *Enzyme and Microbial Technology*, **5**, 383–8.

Dawson, P. S. S. (1974). *Biotechnology and Bioengineering Symposium*, **4**, 809–19.

Debabov, V. G. (1982). In *Overproduction of Microbial Products*, Eds Krumphanzl, V. and Vanek, Z., pp. 345–52. London, Academic Press.

Demain, A. I. (1980). *Search*, **11**, 148–51.

Demain, A. L. (1984). In *Biotechnology of Industrial Antibiotics*. Ed. Vandamme, E. J., pp. 33–44. New York, Marcel Dekker.

Ditchburn, P., Giddings, B. and MacDonald, K. D. (1974). *Journal of Applied Bacteriology*, **37**, 515–23.

Dixon, J. E., Stolzenbach, F. E., Berenson, J. A. and Kaplan, N. O. (1973). *Biochemical and Biophysical Research Communications*, **52**, 905–12.

Dostalek, M., Haggstrom, C. and Molin, N. (1972). *Fermentation Technology Today, Proceedings of the IV International Fermentation Symposium*, pp. 497–511. Ed. Terui, G., Tokyo, Society of Fermentation Technology, Japan.

Douzou, P. and Balny, C. (1977). *Proceedings of the National Academy of Sciences, USA*, **74**, 2297–300.

Dulaney, E. L. and Dulaney, D. D. (1967). *Transactions of the New York Academy of Sciences*, **29**, 792–9.

Dunnill, P. and Rudd, M. (1984) *Biotechnology and British Industry*, Swindon, Science and Engineering Research Council (U.K.).

Eagon, R. G. (1961). *Journal of Bacteriology*, **83**, 736–7.

Egli, T. and Lindley, N. D. (1984). *Journal of General Microbiology*, **130**, 3239–49.

Elander, R. P., Mabe, J. A., Hamill, R. L. and Gorman, M. (1971). *Folia Microbiologica*, **16**, 157–65.

Feren, C. J. and Squires, R. W. (1969). *Biotechnology and Bioengineering*, **11**, 583–92.

Ferenczy, L., Kevei, F. and Zsolt, J. (1974). *Nature (London)*, **248**, 793–4.

Fiechter, A. (1982). In *Advances in Biotechnology, 1. Scientific and Engineering Principles*. Eds, Moo-Young, M., Robinson, C. W. and Vezina, C, pp. 201–67. Oxford, Pergamon Press.

Forniani, G., Leoncini, G., Segin, P., Calabria, G. A. and Dacha, M. (1969). *European Journal of Biochemistry*, **7**, 214–22.

Furbish, J. (1977). *Proceedings of the National Academy of Sciences, USA*, **74**, 3560–3.

Geigert, J., Neidleman, S. L. and Hirano, D. S. (1983). *Carbohydrate Research*, **13**, 159–62.

Glover, D. M. (1984), *Gene Cloning: The Mechanics of DNA Manipulation*. London, Chapman and Hall.

Glover, D. M. (Ed.) (1985). *DNA Cloning: A Practical Approach*. Vols. 1 and 2. Oxford, IRL Press.

Grierson, D. and Covey, S. N. (1984). *Plant Molecular Biology*. Glasgow, Blackie.

Godfrey, T. and Reichert, J. (1983). *Industrial Enzymology*. London, Macmillan.

Goldberg, I., Rock, J. S., Ben-Bassat, A. and Mateles, R. I. (1976). *Biotechnology and Bioengineering*, **18,** 1657–68.

Goldstein, L. (1972). *Biochemistry* **11,** 4072.

Gore, J. H., Reisman, H. B. and Gardner, C. H. (1968). US Patent 3,404,102.

Gray, P. W., Aggarwal, B. B., Benton, C. V., Bringman, T. S., Henzel, W. J., Jarrett, J. A., Leung, D. W., Moffat, B., Ng, P., Svedersky, L. P., Palladino, M. A. and Nedwin, G. E. (1984). *Nature*, **312,** 721–4.

Gregory, G. (1984). *New Scientist*, **104,** 12–15.

Guilbault, G. G. (1980). *Enzyme and Microbial Technology*, **2,** 258–64.

Hamer, G. (1979). In *Economic Microbiology*, Vol. 2. Ed., Rose, A. H., pp. 31–45. London, Academic Press.

Hamilton, B. K., Hsiao, H-Y., Swann, W. E., Anderson, D. M. and Delente, J. J. (1985). *Trends in Biotechnology*, **3,** 64–8.

Hamlyn, P. F. and Ball, C. (1979). In *Genetics of Industrial Microorganisms* Eds, Sebek, O. K. and Laskin, A. I., pp. 185–91. Washington, American Society for Microbiology.

Hansen, E. C. (1896) In *Practical Studies in Fermentation* (Translated by Miller, A. K.) Chapter 1, pp. 1–76. London, Spoon.

Hanson, R. S. (1980). *Advances in Applied Microbiology*, **26,** 3–39.

Harrison, D. E. F. (1973). *Journal of Applied Bacteriology*, **36,** 309–14.

Henley, J. P. and Sadana, A. (1985). *Enzyme and Microbial Technology*, **7,** 50–60.

Hersbach, G. J. M., Van der Beek, C. P. and Van Dijck, P. W. M. (1984). In *Biotechnology of Industrial Antibiotics*. Ed., Vandamme, E. J., pp. 45–140. New York, Marcel Dekker.

Higgins, I. J., Best, D. J. and Hammond, R. C. (1980). *Nature*, **286,** 561–4.

Higgins, I. J., Best, D. J., Hammond, R. C. and Scott, D. (1981). *Microbiological Reviews*, **45,** 556–90.

Hirose, Y. and Shibai, H. (1980). *Advances in Biotechnology*, Vol. 1. Eds, Moo-Young, M., Robinson, C. W. and Vezina, C., pp. 329–33. Toronto, Pergamon Press.

Hooykaas-Van Slogteren, G. M. S., Hooykaas, P. J. J. and Schilperoort, R. A. (1984). *Nature*, **311,** 763–4.

Hopwood, D. A. (1976). In *Microbiology–1976*, Ed, Schlessinger, D., pp. 558–62. Washington, American Society for Microbiology.

Hopwood, D. A. (1979). In *Genetics of Industrial Microorganisms*, Eds, Sebek, O. K. and Laskin, A. I., pp. 1–9. Washington, American Society for Microbiology.

Hopwood, D. A., Malpartida, F., Kieser, H. M., Ikeda, H., Duncan, J., Fujii, I., Rudd, B. A. M., Floss, H. G. and Omura, S. (1985). *Nature*, **314,** 642–4.

Huber, F. M. and Tietz, A. J. (1984). In *Biotechnology of Industrial Antibiotics*. Ed. Vandamme, E. J., pp. 551–68. New York, Marcel Dekker.

Hue, D., Corvol, P., Menard, J. and Sicard, P. J. (1976). *Process Biochemistry*, **July/August,** 20–4.

Hustedt, H., Kroner, K. H., Stach, W. and Kula, M-R. (1978). *Biotechnology and Bioengineering*, **20,** 1989–2005.

Ichikawa, T., Date, M., Ishikura, T. and Ozaki, A. (1971). *Folia Microbiologica*, **16,** 218–24.

Jones, J. B. and Taylor, K. E. (1976). *Canadian Journal of Chemistry*, **54,** 2974–80.

Karube, I. (1984). In *Biotechnology and Genetic Engineering Reviews*, Vol. 2. Ed. Russell, G. E., pp. 313–41. Newcastle, Intercept.

Kase, H. and Nakayama, K. (1972). *Agricultural and Biological Chemistry*, **36,** 1611–21.

Kinoshita, S., Udaka, S. and Shimono, M. (1957). *Journal of General Applied Microbiology*, **3,** 193–205.

Klibanov, A. M., Nathan, N. O. and Kamen, M. D. (1978). *Proceedings of the National Academy of Sciences, USA*, **75,** 3640–3.

Komatsu, K., Mizumo, M. and Kodaira, R. (1975). *Journal of Antibiotics*, **28,** 881–7.

Kornberg, H. L. (1972). In *Essays in Biochemistry*, Vol. 2, London, Academic Press.

Koukolikova-Nicola, Z., Shillito, R. D., Hohn, B., Wang, K., Van Montagu, M. and Zembryski, P. (1985). *Nature*, **313,** 191–196.

Kula, M-R. (1979). In: *Applied Biochemistry and Bioengineering*, Vol. 2. Eds Wingard, L. B. Katchalski-Katzir, E. and Goldstein, L. London, Academic Press.

Kurth, R. and Demain, A. L. (1984). In *Biotechnology of Industrial Antibiotics* Ed. Vandamme, E. J., pp. 781–90. New York, Marcel Dekker.

Kyowa Hakko Kogyou (1983). European patent application 73062.

Large, P. J. (1981). In *Microbial Growth on C1 Compounds*. Eds Dalton, H. pp. 55–69. London, Heyden and Son.

Large, P. J. (1983). *Methylotrophy and Methanogenesis*, pp. 1–88. Wokingham, Van Nostrand.

Lehninger, A. L. (1982). *Principles of Biochemistry*, pp. 331–680. New York, Worth Publishers.

LeGrys, G. A. and Solomon, G. L. (1977). Patent application 23128.

Lewin, B. (1983). *Genes*. New York, John Wiley.

Lowe, C. R. (1981). In *Topics in Enzyme and Fermentation Biotechnology*, Vol. 5. Ed A. Wiseman, pp. 13–146. Chichester, Ellis Horwood.

MacDonald, K. D. and Holt, G. (1976). *Science Progress*, **63,** 547–73.

Mainwaring, W. I. P., Parish, J. H., Pickering, J. D. and Mann, N. H. (1982). *Nucleic Acid Biochemistry and Molecular Biology*. Oxford, Blackwell.

Malik, V. S. (1980). *Trends in Biochemical Sciences*, **March. 5,** (3), pp. 68–72.

Malik, V. (1982). *Advances in Applied Microbiology*, **28,** 28–116.

Maniatis, T., Fritsch, E. F. and Sambrook, J. (1982). *Molecular Cloning: A Laboratory Manual*. Cold Spring Harbor, USA.

Mansson, M-O., Larsson, P-O. and Mosbach, K. (1978). *European Journal of Biochemistry*, **86,** 455–61.

Mansson, M-O., Larsson, P-O. and Mosbach, K. (1979). *FEBS Letters*, **98,** 309–13.

Mantell, S. H., Matthews, J. A. and McKee, R. A. (1985). *Principles of Plant Biotechnology*, pp. 191–6. Oxford, Blackwell.

Marsh, R. A. and Pinkney, J. T. (1985). In *The World Biotech Report, 1985. Vol. 1: Europe. Proceedings of Biotech '85, Europe, Geneva*. pp. 253–62. London, Online Publications.

Martin, J. F., Gill, J. A., Naharro, G., Liras, P. and Villanueva, J. R. (1979). In *Genetics of Industrial Microorganisms*. Eds. Sebek, O. K. and Laskin, A. I., pp. 205–209. Washington, American Society for Microbiology.

Martinek, K. and Berezin, I. V. (1977). *Journal of Solid Phase Biochemistry*, **2,** 350–1.

Martinek, K. and Semenov, A. N. (1981). *Journal of Applied Biochemistry*, **3,** 93–126.

Martinek, K., Klibanov, A. M., Goldmacher, K. S. and Berezin, I. V. (1977a). *Biochimica Biophysica Acta*, **485,** 1–12.

Martinek, K., Klibanov, A. M., Tchernyshera, A. V., Mozhaev, V. V., Berezin, I. V. and Glotov, B. O. (1977b). *Biochimica Biophysica Acta*, **485,** 13–28.

Marutzky, P., Petessen-Barstel, H., Elossdorf, J. and Kula, M-R. (1974). *Biotechnology and Bioengineering*, **16,** 1449–58.

Mateles, R. J. (1979). *Society of General Microbiology Symposium*, **29,** *Microbial Technology: Current State and Future Prospects*, Eds Bull, A. T., Ellwood, D. C. and Ratledge, C., pp. 29–52. Cambridge, Cambridge University Press.

McNerney, T. and O'Connor, M. L. (1980). *Applied and Enironmental Microbiology*, **40,** 370–5.

Meyer, O. and Schlegel, H. G. (1983). *Annual Reviews of Microbiology*, **37,** 277–310.

Miwa, K., Nakamori, S. and Mimose, H. (1981). *Abstracts of 13th International Congress of Microbiology (Boston, USA)*. p. 96.

Monod, J. (1942). *Recherches sur les Croissances des Cultures Bacteriennes*, 2nd edition. Paris, Hermann and Cie.

Mozhaev, V. V. and Martinek, K. (1984). *Enzyme and Microbial Technology*, **6,** 50–60.

Merrick, M. and Dixon, R. (1984). *Trends in Biotechnology*, **2,** 162–6.

Murai, N., Sutton, D. W., Murray, M. G., Slightom, J. L., Merlo, D. J., Reichert, N. A., Sengupta-Gopalan, Stock, C. A., Barker, R. F., Kemp, J. D. and Hall, T. C. (1983). *Science*, **222,** 476–82.

Murrell, J. C. and Dalton, H. (1983). *Journal of General Microbiology*, **129,** 3481–6.

Nakao, Y., Kanamaru, T., Kikuchi, M. and Yamatodani, S. (1973). *Agricultural and Biological Chemistry*, **37,** 2399–404.

Nakayama, K., Kituda, S. and Kinoshita, S. (1961). *Journal of General and Applied Microbiology*, **7**, 41–51.

Nathan, L. (1930). *Journal of the Institute of Brewing*, **36**, 538–50.

Neidleman, S. L. and Geigert, J. (1983). *Trends in Biotechnology*, **1**, 21–25.

Neidleman, S. L., Amon, W. F. and Geigert, J. (1981). US Patent 4,247,641.

Okachi, R. and Nara, T. (1984). In *Biotechnology of Industrial Antibiotics*, pp. 329–66. Ed. Vandamme, E. J. New York, Marcel Dekker.

Old, R. W. and Primrose, S. B. (1985). *Principles of Gene Manipulation: An Introduction to Genetic Engineering*, 3rd edn Oxford, Blackwell.

Ohlson, S., Flygave, S., Larsson, P-O., Mosbach, K. (1980). *European Journal of Applied Microbiology*, **10**, 1–9.

Ozias-Akins, P. and Lorz, H. (1984). *Trends in Biotechnology*, **2**, 119–23.

Pal, P. K. and Gertler, M. M. (1983). *Thrombosis Research*, **29**, 175–85.

Palmer, T. (1981). *Understanding Enzymes*. Chichester, Ellis Horwood.

Pastore, M. and Morisi, F. (1976). In *Methods in Enzymology*, Vol. 44. pp. 822–30. Ed. Mosbach, K. New York, Academic Press.

Paszkowski, J., Shillito, R. D., Saul, M., Mandak, V., Hohn, T., Hohn, B. and Potrykus, I. (1984). *EMBO Journal*, **3**, 2717–22.

Pennica, D., Nedwin, G. E., Hayflick, J. S., Seeburg, P. H., Derynck, R., Palladino, M. A., Kohr, W. J., Aggarwal, B. B. and Goeddel, D. V. (1984). *Nature*, **312**, 724–9.

Pirt, S. J. (1973). *Journal of General Microbiology*, **75**, 245.

Pirt, S. J. (1975). *Principles of Microbe and Cell Cultivation*. Oxford, Blackwell.

Pirt, S. J. and Kurowski, W. M. (1970). *Journal of General Microbiology*, **63**, 357–66.

Podojil, M., Blumaverova, M. Culik, K. and Vanek, Z. (1984). In *Biotechnology of Industrial Antibiotics*, pp. 259–80. Ed. Vandamme, E. J. New York, Marcel Dekker.

Pontecorvo, G., Roper, J. A., Hemmons, L. M., MacDonald, K. D. and Bufton, A. W. J. (1953). *Advances in Genetics*, **5**, 141–238.

Poulsen, P. B. (1984). In *Biotechnology and Genetic Engineering Reviews*, Vol. 1. Ed. Russell, G. E. pp. 121–40. Newcastle, Intercept.

Quayle, J. R. (1980). *Biochemical Society Transactions*, **8**, 1–10.

Queener, S. and Swartz, R. (1979). In *Economic Microbiology 3: Secondary Products of Metabolism* Ed. Rose, A. H. pp. 35–123. London, Academic Press.

Ray Chowdhurry, M. K., Goswani, R. and Chakrabarti, P. (1980). *Analytical Biochemistry*, **108**, 126–8.

Reed, G. and Peppler, H. J. (1973). *Yeast Technology*, pp. 664–8. Westport, Connecticut, Avi.

Rehacek, J. and Schaefer, J. (1977). *Biotechnology Bioengineering*, **19**, 1523–34.

Rivierre, J. (1977). *Industrial Applications of Microbiology* (Trans. and ed Moss, M. O. and Smith, J. E.). Brighton, Surrey University Press.

Robers, J. F. and Floss, H. G. (1970). *Journal of Pharmacological Science*, **59**, 702–3.

Robinson, P. J., Wheatley, M. A., Janson, J. C., Dunnil, P. and Lilly, M. D. (1974). *Biotechnology and Bioengineering*, **16**, 1103–112.

Saltero, F. V. and Johnson, M. J. (1954). *Applied Microbiology*, **2**, 41–4.

Sano, K. and Shiio, I. (1970). *Journal of General Applied Microbiology*, **16**, 373–91.

Schell, J. and Van Montagu, M. (1983). *Biotechnology*, **1**, 175–80.

Schmidt, R. D. (1979). *Advances in Biochemical Engineering*, **12**, 41–118.

Schulze, K. L. and Lipe, R. S. (1964). *Archiv für Mikrobiologie*, **48**, 1–20.

Schutte, H., Flossdorf, J., Sahn, H. and Kula, M-R. (1976). *European Journal of Biochemistry*, **62**, 151–160.

Sermonti, G. (1969). *Genetics of Antibiotic Producing Organisms*. London, Wiley-Interscience.

Shehata, T. E. and Narr, A. G. (1971). *Journal of Bacteriology*, **107**, 210–25.

Shepard, J. F., Bidney, D., Barsby, T. and Kemble, R. (1983). *Science*, **219**, 683–88.

Simon, L. M., Szelei, J., Szajáni, B., and Boross, L. (1985). *Enzyme and Microbial Technology*, **7**, 357–60.

Sinclair, C. G. and Kristiansen, B. (1987). In *Fermentation Kinetics and Modelling*, Ed. Bu'Lock, J. D. Open University Press in association with Paradign Press.

Sipiczki, M. and Ferenczy, L. (1977). *Molecular and General Genetics*, **157**, 77–83.

Smith, S. R. L. (1980). *Philosophical Transactions of the Royal Society* (*London*) B, **290**, 341–54.

Smith, G. M. and Calam, C. T. (1980). *Biotechnology Letters*, **2**, 261–6.

Stanbury, P. F. and Whitaker, A. (1984). *Principles of Fermentation Technology*. Oxford, Pergamon Press.

Stanley, S. H., Prior, S. D., Leak, D. J. and Dalton, H. (1983). *Biotechnology Letters*, **5**, 487–92.

Stoppok, E., Schomer, U., Segner, A., Matel, H. and Wagner, F. (1980). Poster presented at the 6th International Fermentation Symposium, London, Ontario, Canada.

Stryer, L. (1981). *Biochemistry*, pp. 235–356. San Francisco, W. H. Freeman and Co.

Szwajer, E., Brodelius, P. and Mosbach, K. (1982). *Enzyme and Microbial Technology*, **4**, 409–13.

Takahashi, K., Nishimura, H., Yoshimoto, T., Okada, M., Ajima, A., Matsushima, A., Tamaura, Y., Saito, Y. and Inada, Y. (1984). *Biotechnology Letters* **6(12)**, 765–70.

Takata, I., Kayashima, K., Tosa, T. and Chibata, I. (1982). *Journal of Fermentation Technology*, **60**, 431–7.

Taylor, I. J. and Senior, P. J. (1978). *Endeavour*, **2**, 31–4.

Tonge, G. W. (1980), *Enzyme and Microbial Technology*, **2(4)**, 342–50.

Torchillin, V. P., Maksimenko, A. V., Smirnov, V. N., Berezin, I. V., Klibanov, A. M. and Martinek, K. (1977). *Biochimica Biophysica Acta*, **522**, 277–83.

Tosaka, O., Karasawa, M., Ikeda, S. and Yoshii, H. (1982). *Abstracts of 4th International Symposium on Genetics of Industrial Microorganisms*, p. 61.

Trevan, M. D. (1980). *Immobilised Enzymes. An Introduction and Application in Biotechnology*. Chichester, John Wiley.

Trevan, M. D. and Groves, S. (1979). *Biochem. Soc. Trans.* **7**, 28–30.

Trilli, A., Michelin, V., Mantovani, V. and Pirt, S. J. (1978). *Antimicrobial Agents and Chemotherapy*, **13**, 7–13.

Trinci, A. P. J. (1969). *Journal of General Microbiology*, **57**, 11–24.

Vallee, B. L. (1980). *Carlsberg Research Communications*, **45**, 423–41.

Vane, J. and Cuatrecasas, P. (1984). *Nature*, **312**, 303–5.

Vezina, C. and Singh, K. (1975). In *The Filamentous Fungi*, Vol. 1, pp. 158–92. Eds, Smith, J. E. and Berry, D. R. London, Arnold.

Vezina, C., Singh, K. and Sehgal, S. N. (1965). *Mycologia*, **57**, 722–36.

Vogel, G. D., Van der Drift, C., Stumm, C. K., Keltjens, J. T. M. and Zwart, K. B. (1984). *Antonie van Leeuwenhoek*, **50**, 557–67.

Walker, J. M. and Gaastra, W. (Eds) (1983). *Techniques in molecular biology*. London. Croom Helm.

Wandrey, C. (1984). In *Biotech Europe 84*, pp. 391–404. Online Publications, Pinner.

Wharton, C. W., Crook, E. M., and Brokelhurst, K. (1968). *European Journal of Biochemistry*, **6**, 572.

Whitaker, A. and Long, P. A. (1973). *Process Biochemistry*, **8**, 27–31.

Wilkinson, A. J., Fersht, A. R., Blow, D. M., Carter, P. and Winter, G. (1984). *Nature*, **307**, 187–8.

Windass, J. D., Worsey, M. J., Pioli, E. M., Pioli, D., Barth, P. T., Atherton, K. T., Dart, E. C., Byrom, D., Powell, K. and Senior, P. J. (1980). *Nature*, **287**, 396–401.

Wingard, L. B., Shaw, C. H. and Castuer, J. F. (1982). *Enzyme and Microbial Technology*, **4**, 137–42.

Wingard, Jr. L. B., Roach, R. P., Miyawaki, O., Egler, K. A. and Klinzing, G. E., (1985). *Enzyme and Microbial Technology*, **7**, 503–9.

Winter, G. and Fersht, A. R. (1984). *Trends in Biotechnology*, **2**, 115–19.

Wodzinski, R. S. and Johnson, M. J. (1968). *Applied Microbiology*, **16**, 1886–91.

Woodward, J. (1985) (Ed.) *Immobilized Cells and Enzymes*. Oxford, IRL Press.

Wyatt, J. M. (1984). *Trends in Biochemical Sciences*, **8**, 19–23.

Zaborsky, O. R. (1972). In *Enzyme Engineering*, Vol. 1. Ed. Wingard, L. B., pp. 211–17. New York, Plenum.

Zaborsky, O. R. (1974). In *Enzyme Engineering*, Vol. 2. Eds. Pye, E. K. and Wingard, L. B., pp. 115–22. New York, Plenum.

Zaks, A. and Klibanov, A. M. (1984). *Science*, **224**, 1249–51.

Zeikus, J. G. (1983). *Advances in Microbial Physiology*, **24**, 215–93.

Zobell, C. E. (1950). *Advances in Enzymology*, **10**, 433–68.

# Index